The Maze of Ingenuity

609

GRE
ROYAL COUNTY
OF BERKSHIRE
LIBRARY AND
INFORMATION SERVICE

READING CENTRAL LIBRARY
ABBEY SQUARE
READING, BERKS RG1 3BQ
TEL: 0118 901 5950

KT-168-586

02. AUG
02 SEP
18. MAR 04
10. NOV
17. MAY 97

026266075X 0 992

The Maze of Ingenuity

Ideas and Idealism in the Development of Technology

Second Edition

Arnold Pacey

The MIT Press
Cambridge, Massachusetts
London, England

© Arnold Pacey 1974, 1992

All rights reserved. No part of this book may be reproduced in any form by any electronic or mechanical means (including photocopying, recording, or information storage and retrieval) without permission in writing from the publisher.

This book was set in Baskerville by DEKR Corporation and printed and bound in the United States of America.

Library of Congress Cataloging-in-Publication Data

Pacey, Arnold.
The maze of ingenuity : ideas and idealism in the development of technology / Arnold Pacey. — 2nd ed.
p. cm.
Includes bibliographical references and index.
ISBN 0-262-16128-1. — ISBN 0-262-66075-X (pbk.)
1. Technology—History. 2. Technology—Social aspects.
I. Title.
T15.P35 1992
609—dc20 91-27441
CIP

Contents

Preface

The function of technology in society is mainly to provide our food, clothing, shelter, defense, and transportation, but creativity in technology, and the urge to innovate, is often driven by quite different considerations—by visions of new worlds, ideals of rational order, and sheer fascination with machines, materials, or inanimate power. Such creativity, essentially idealistic, commands respect both for its achievements and for the dedication of those through whom it works most strongly. Yet it is clearly a dangerous force unless constrained by response to social need, ethical principles, or economic disciplines.

In another book, *The Culture of Technology,* I discussed some of the problems of making modern technology responsive to social needs. My concern was that the idealistic impulse associated with creativity tends to divert effort away from this aim so that there is only limited *social control* of technology, even in democratic societies. In the present book, my purpose is to discuss historical rather than modern situations, and many of the examples will show inventive and skilled people evading *economic constraints* by allying themselves with idealistic ventures—such as cathedral building, or more recently, scientific research—and by seeking out institutions that can insulate their work from market pressures.

The original edition of the book attempted to deal with another theme as well, namely interactions between western and Asian technologies. I have recently been able to develop that topic in a separate volume, *Technology in World Civilization,* leaving scope for this new edition of *The Maze of Ingenuity* to concentrate wholly on idealistic trends in western technology. Although either of the books can be read independently, they are intended to complement one another, and they cover approximately the same long period in history, from the Middle Ages to the present—a period chosen because of the marked expansion in the use of machines (especially water mills) that occurred in the early Middle Ages, not only in Europe but also in much of Asia.

The two books therefore represent two ways of looking at a single theme. Many other interpretations of the history of technology are possible, some of which would stress economic circumstances more than I do, including such developments as population growth, the expansion of commerce, and the rise of capitalism. I have no wish to deny the importance of these influences, but the purpose of this book is to show that non-economic, "idealistic" influences were also significant.

One problem with the history of technology is that there is no single theoretical perspective or "paradigm" in this field, and no integrated view of the whole. Thus apart from books that look at technological development in relation to the working of the economy (e.g., Landes, Rosenberg), there are histories of technology that examine the military aspect (McNeill), and there are essays that describe the role of aesthetic impulses in technological innovation (Hindle, C. S. Smith) or vision, myth, and fantasy (Basalla, Winner). Among all these, Basalla comes nearest to showing how economic factors and non-economic impulses may be related.

The idealistic trends in technology with which this book is concerned include the aesthetic impulses and the visionary thinking that some of these authors stress, but I am also concerned with the ways in which many inventions have arisen from discoveries in science. Both art and science involve exploration of the material world in search of meaning and significance—meaning related to rational understanding of nature (in science), and related to sense experience (in art). There is every reason to think that technically inventive activity is also an exploration related to both these goals, but with deeper involvement in some kinds of experience—especially experience of motion, energy, and materials.

Projects involving exploration of this kind are classified here as involving the pursuit of "technical ideals"—that is, ideals centered on fascination with technical matters and with virtuosity in the use of machines and techniques (hence "virtuosity values" in *The Culture of Technology*). Broadly speaking, technical ideals can be regarded as being of two main kinds, related either to the intellectual, rational, or scientific content of technology ("intellectual ideals"), or to sense experience and aesthetic satisfactions associated with building and construction, working metals, controlling engines, or flying aircraft.

Beyond direct experience of technical matters, whether intellectual or aesthetic, there are ideals concerned with the wider meaning and purpose of technology. Here we may contrast the social ideals of those who wish to concentrate on providing for human need with the quite different approach of people who perceive technology in terms of power. The latter may be thinking of political and military power, or of a Faustian program for mastery over nature. Either attitude is associated with a tendency to create monuments and symbols in which technique is pushed beyond what is justified by strict practicality. One way of describing this sort of idealism in technology, therefore, is to say that it is related to the creation of symbols, as we shall see with regard to cathedrals and clocks (chapters 1 and 2), as well as bridges (chapter 8) and "high" technology.

The first six chapters of this version of the book remain much as they were in the original edition, though with minor revisions. Chapter 7 covers the same ground as before, but has been entirely rewritten to incorporate new insights and to take in some material formerly in chapter 8. The final three chapters are very largely new to this edition and attempt to give a more rounded view of nineteenth-century technology than was previously possible, and to sketch some idealistic trends in twentieth-century innovation.

The first edition of the book was written while I was teaching the history of science and technology at the University of Manchester Institute of Science and Technology (UMIST), along with one course at the University of Leeds (in 1969–70). Work on this new edition has been carried on in parallel with tutorial teaching for an Open University course entitled "Technology and Change", dealing with the history of nineteenth-century industry. Both kinds of teaching experience have contributed to the book, and several students are acknowledged in the notes. I have also profited from readers' reactions to the first edition, some of the most helpful coming from Angus Buchanan and Cyril Stanley Smith. Among colleagues and others to whom I owe specific debts are Donald Cardwell, Chris Crosland, the late Wilfred Farrar, and (no relation) David Farrar (who drew some of the diagrams); also Lisa Barker, Dinie Blake, Richard Hills, Stuart Smith, and Charles Webster. Jerry Ravetz read chapters and discussed ideas connected with both editions. My brother Philip Pacey has helped extensively with library work for this new edition, and with Gill Pacey helped assemble material for illustrating the original version.

I am also indebted to the following for permission to reproduce illustrations:

Birmingham Public Libraries (figure 35)

Cambridge University Press (figure 23)

Dover Publications Inc., New York (figures 19, 20, and 25)

Editorial Gustavo Gili, Barcelona (figure 17)

Faber and Faber Ltd., London (figure 8)

David Fulton (Publishers), Ltd., London (figure 41)

Harvard University, Fogg Art Museum (figure 16)

Ironbridge Gorge Museum Trust (figures 37, 38, and 39)

Liverpool University Press (figure 14)

The Newcomen Society, London (figure 29)

Penguin Books Ltd. (figure 5)

Arnold Pacey
Addingham, February 1991

The Maze of Ingenuity

1

The Cathedral Builders: European Technical Achievement between 1100 and 1280

Introduction: Innovations of the 1090s

In the year 1093, foundations were laid for a new cathedral at Durham, on a high, river-bound site in northern England. The building was being erected by a French bishop who had lived at Rouen and near Le Mans before being appointed to Durham, and its construction combined some of the best technical detail of recent French buildings with a number of important innovations.

The building of this cathedral shows much evidence of the efficiency of the new Norman rulers of England and of their growing technical competence. But in England and France at this time, material civilization and technical skills were still greatly inferior to those that could be found at the same date in some of the Islamic lands around the Mediterranean and farther afield, in China and India. There had certainly been significant technical development in England, as shown by the 6,000 water mills that the Normans noted in 1086 during their "Domesday" survey. But these mills were very simple structures involving much less skill in the use of water resources than the dams and irrigation systems that Islamic builders had been constructing in Spain at about the same time.[1] And nowhere in northern Europe was there knowledge of complex machines, or of medicine and chemistry, such as could be found at major Islamic centers of learning, especially Baghdad.

It is true that a thousand years earlier, when the Roman Empire extended from the Mediterranean into France and Britain, highly developed engineering skills had been used in this region—now largely forgotten. However, the period that encompasses the Domesday survey and the beginnings of Durham cathedral is a landmark period in the history of western technology for several reasons. First, there was the beginning of a rediscovery of Roman and Greek technical knowledge. Second, much was learned from this time onward by contact with the Islamic civilization and the eastern countries with which it was in touch. Most important of all was the energy and idealism, the determination and imagination, with which Europeans transformed what they learned from other civilizations, creating a new and distinctive technology.

Islamic rule had extended from the lands to the east and south of the Mediterranean into Sicily and Spain, but during the eleventh century, Christian kings reconquered some of these territories and organized a crusade to liberate Jerusalem. One result was that western Europeans had opportunities to observe Islamic techniques at close quarters. The most important contacts were made in Spain, especially after Toledo was captured by Christian forces in 1085 (figure 1). Islamic scholars living in that city possessed books by Greek and Roman authors in Arabic translation. They had books on new techniques in mathematics also, some based on the use of Indian "Hindu" numerals. They had also experimented with clocks driven by water, some containing sophisticated gear trains that were discussed in original writings on mechanics. Soon Toledo became a center for translating books of these various kinds into Latin, the language of European learning.

Europe's debt to Islamic civilization, to India, and indeed to China, has been discussed in a separate but complementary volume.[2] The purpose of the present book is to examine the imagination, determination, and vision with which Europeans developed the techniques they either borrowed from other civilizations or invented themselves. A foretaste of this is offered by remarkable developments in shipbuilding in Scandinavia some time before AD 900. The great clinker-built boats of the Norsemen or Vikings have been described as "the major technical achievement of the age before the . . . cathedrals."[3] We may also wonder at the boldness and skill with which these ships were sailed across the North Atlantic to a colony in Greenland, founded about AD 985, and from there to parts of the North American coastline.

The Norman families who ruled England a century after this were of Viking descent, so we should not be surprised by their energy or by the vigor of their rebuilding program at Durham and other centers. But the characteristic attitude of western

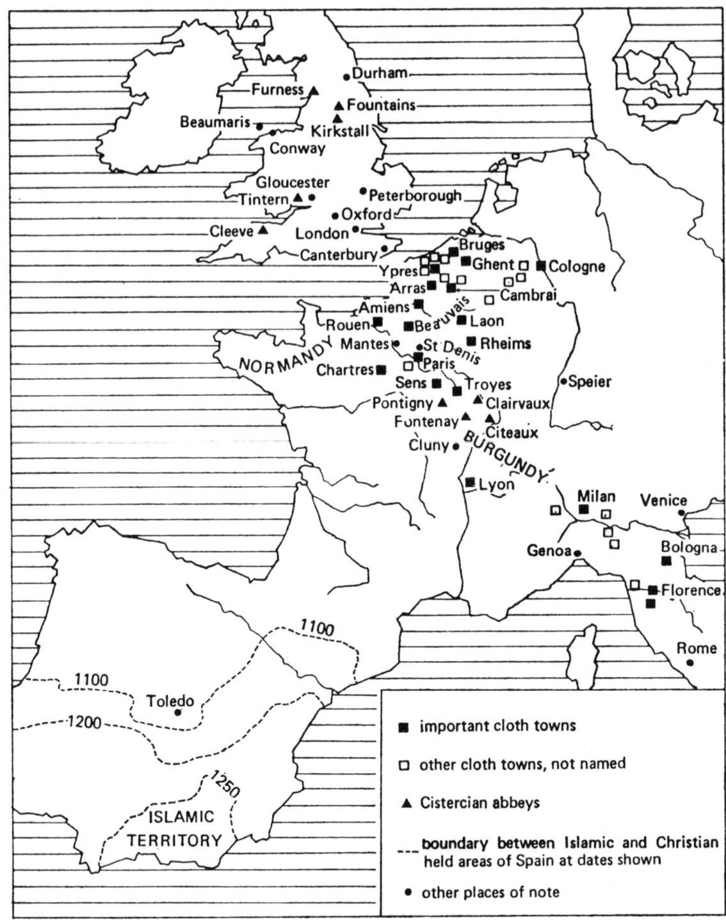

Figure 1
Places mentioned in chapter 1, showing the distribution of major cloth towns.

Europeans to the development of new techniques had other roots as well, some in the abbeys and monasteries that were being founded all over Europe, and some in the commerce of the "cloth towns" (figure 1). And we may perhaps note two aspects of this approach to technology, one exploitative, in a political and social sense, and the other quite markedly idealistic.

The exploitative approach can be illustrated by the way water mills in England were operated by the abbeys or manors that owned them. In many instances, peasants were compelled to give up their hand mills and grind their corn at the lord's or the abbey's mill. This made people more dependent on the lord or abbot, and at the same time provided him with a source of income as they paid over a proportion of their grain to the miller. Another example is the way in which church authorities used cathedral-building schemes to dominate life in the towns where the cathedrals were built, and to retain the loyalty of citizens and skilled artisans. In these respects, machines and construction projects were used to reinforce the wealth and power of established authorities, and some commentators have seen this use of technology for "social control" as particularly significant.[4]

However, to understand the creativity of craftsmen and technicians, we must also notice a different strand in the effort that went into the building of abbeys and cathedrals. In other civilizations, the construction of temples and monasteries quite frequently involved ambitious, technically challenging work. In China at this very date, Buddhist monasteries were making innovative use of cast iron for building components and bells, and parallels could be drawn from other cultures. In nearly all such projects, while the temple was certainly a means of asserting the role of some leading group in society, it was at the same time a work of imagination, setting forth an ideal of strength and permanence in contrast to the frailty and feebleness of human beings. It was in striving for these qualities, often through building on a massive scale but sometimes also through refinement of technique, that many of the challenges came.

In Europe, these attitudes are exemplified by a remarkable development of building techniques beginning just before 1100. In many respects, building methods were refined far beyond the point that was necessary in practical terms solely for strength and permanence, as if beyond these qualities there was a *technical ideal.* And the pursuit of such ideals, not only in building but also in mechanical design, reflects a stirring of the imagination that enabled Europeans at about this date to grasp the immense possibilities of technical progress in a way the Romans never had, and the Chinese had perhaps only momentarily.

Fundamental to all this, however, and making possible the changes of the twelfth and thirteenth centuries, was the underlying prosperity of the Norman period. Much of it was related to the numerous water mills noted in the Domesday survey and to such agricultural techniques as the rotation of crops and the use of a heavy wheeled plow that had been inherited from earlier centuries. And it is perhaps significant that agriculture and the use of water mills had already by 1100 excelled most Roman practice. But of course, with other practical arts where the Romans had achieved more, Europe had fallen well below their achievement, and much had been lost. Building was the technology that the Romans had developed most fully, with their aqueducts and their enormous vaulted public buildings. The secret of much of their success was the way in which they used concrete in conjunction with masonry construction. This important technique had since been lost and was not entirely recovered until about 1800. And this brings us back to the cathedral at Durham, for the technical problems faced by the builders there were ones that the Romans could have solved using concrete but for which the Norman builders had to develop entirely different solutions. And we may note that, apart from the lack of any knowledge of Roman concrete, there were great problems about making a satisfactory mortar that continued to trouble builders throughout the Middle Ages.

The problem at Durham was to provide a high, wide church with a stone ceiling or vault. This was, indeed, the great technical challenge in European architecture at the time. Islamic builders who were faced with the need to accommodate a large number of people in a mosque would often construct a dome. But not knowing how to create a dome, medieval Europeans concentrated on long, narrow buildings with a basilican, nave-and-aisles plan and tackled the problem of spanning the relatively narrow nave with a stone vault.

To make a vault rather than being content with a timber ceiling was desirable partly as a protection against fire, but the vault was also designed to provide a ceiling of the same texture and color as the walls below. To the eye, timber roof beams left exposed appeared to keep the walls apart, but a stone vault, springing from the walls and curving across between them, gave the whole building a more unified appearance. However, the problem of building such a vault over any space as wide as the nave of Durham Cathedral was something that had not been attempted in medieval Europe before about 1075. Then, in France and the Rhineland, a number of churches were given simple vaulted ceilings that offered a variety of tentative solutions to the problem. Almost all these vaults were conceived as continuous stone shells forming a smoothly curved ceiling to the church. The innovation at Durham consisted in subdividing the vault into small areas by means of stone ribs. These ribs were really arches, crossing the church both transversely and diagonally; and in order that they should all reach up to the same height, the transverse arches rose sharply to a point, whereas the diagonal ones were semicircular (see figures 2 and 3). This stage in the development of Durham was not reached until the 1120s. But even then, pointed arches were almost unknown except in a few Cistercian abbey churches, although later in the twelfth century they became the most characteristic feature of "Gothic" architecture.

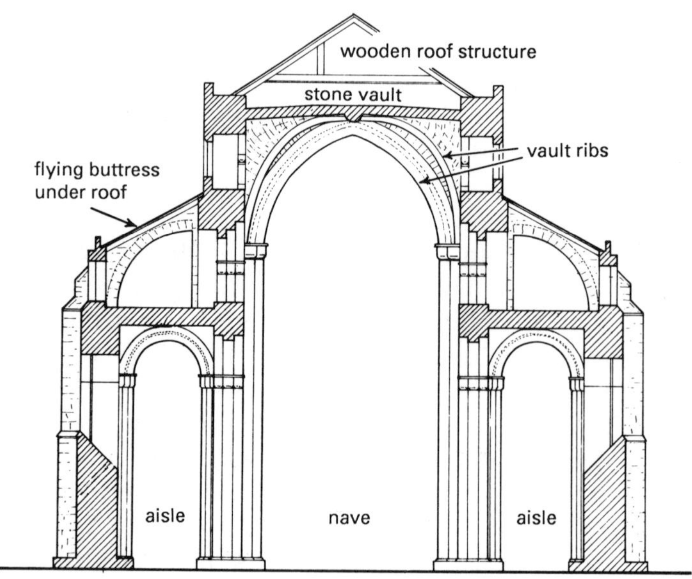

Figure 2
Durham Cathedral: cross-section of the nave. The cathedral was begun in 1093; the nave was built c.1104–30.

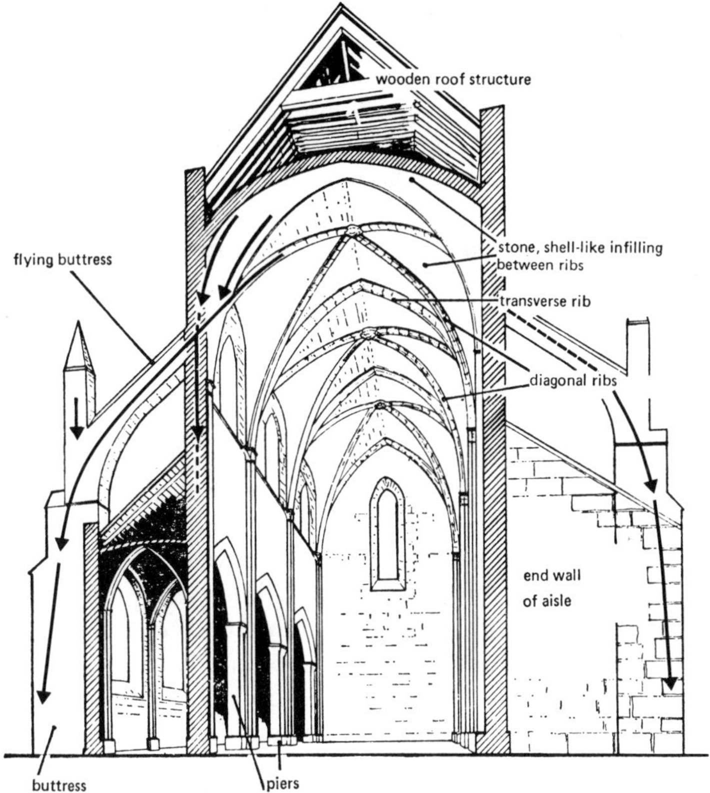

Figure 3
The principle of the rib vault: arrows and dashed lines show how thrusts (forces) from the ribs are carried partly by the flying buttresses.

The ribs in the Durham vaults had both a technical and an aesthetic function. Thus they made possible a stronger structure, which was also lighter in weight, and at the same time, because the ribs were built first and the rest of the vault was filled in later, they made the whole thing easier to construct, involving the use of less centering and scaffolding. The aesthetic improvement obtained by using the rib vault was that its surface, patterned with ribs, could be made to match the piers and walls more closely; stone shafts attached to the piers ran from floor level to the vault, and the ribs were made to continue their vertical line into its curving shape.

The introduction of rib vaults at Durham gave rise to one major structural problem. The forces resulting from the weight of the vaults could not simply be supported by the piers and walls, because they did not act downward, but rather they exerted an outward thrust, tending to push the walls apart. In addition, these forces were concentrated into a very small area at the convergence of each group of ribs. It was necessary, therefore, to buttress the piers high up, at the level where the walls had to withstand these forces. To do this by building conventional solid buttresses against the piers would involve blocking the aisle of the church, however. So instead, a buttress was built against the aisle wall, and a leaning, bridgelike piece of masonry was made to span across the aisle to the point where the vault thrusts needed to be counteracted; and this device constituted one of the earliest flying buttresses in western Europe (figure 2).

The three technical innovations at Durham, then, were the rib vault, the pointed arch, and the flying buttress. Their introduction served well to illustrate the medieval craftsman's skill and his capacity for change; it is also instructive to note how the new techniques were worked out with a knowledge of recent continental buildings (e.g., Speier Cathedral on the Rhine), and perhaps even drew on knowledge of the pointed arches used by Islamic builders in Spain. The bishop who began the building work was perhaps also responsible for assembling master craftsmen with this knowledge, for he had just previously spent three years traveling on the continent.

The question arises as to whether other practical arts besides building possessed this capacity for innovation. Building is the one medieval art of which there are extensive remains and good documentation, but it seems likely that if more were known about contemporary machines, particularly water mills, a parallel development would be noted—that is, one would be able to discern a fairly rapid improvement in design and utilization from the middle of the eleventh century. The Domesday Book gives some evidence of such innovations, although much of it is difficult to interpret. Fulling mills, for example (used for the finishing process on woollen cloth), may possibly be referred to there; they were certainly used in the twelth century and played a very important part in the textile trade from the thirteenth century onward.

The skills necessary for the building of machines developed in parallel with the art of building if only because, to some degree, the same people were responsible for both. The architect of a cathedral would be a master craftsman whose job included not only the design of the building but also such work in mechanics as the construction of hoists for lifting heavy stones to great heights. Detailed evidence of this is lacking for the builders of Durham, but eighty years later, William of Sens, architect of Canterbury Cathedral, impressed the monks there by making, as they wrote, "the most ingenious machines for loading and unloading ships, and for drawing the mortar and stones."[5] Judging by the size of stones used high up in the cathedral walls, it seems that William's hoists had a capacity of about half a ton. Such a machine would be worked by a windlass, but there was another type of hoist used by the builders for lifting the stones of the vault. This was the "Great Wheel," supported on the roof beams above the vault and powered by a treadmill.

Another master mason known for his interest in mechanical subjects was Villard de Honnecourt. He is now famous for the notebook he filled with sketches, which included a waterwheel driving a saw, a screw jack for lifting heavy weights, and a mechanical eagle that turned its head. Villard's notebook dates from about 1235, but it seems highly probable that master builders of about 1100 would have the same breadth of interests. And since in many places where they worked, such men would be the most expert craftsmen available, they must sometimes have been called on for advice about the construction or repair of a local mill.

Practical Arts and the Monastic Orders

After 1100 the improvement of water mills was perhaps most actively pursued by the monks and lay brothers belonging to the abbeys and priories that proliferated at this time. In these institutions, the arts of building and milling, agriculture and water supply were carried on side by side, and the organization that linked the abbeys of each order proved to be an effective means of communicating technical ideas and practical skills from one part of Europe to another.

The pattern of life within most of the great abbeys was based on the program laid down by St. Benedict about AD 530. But with the passage of time, there was a tendency for the monks to lose the enthusiasm and relax the austerity for which their predecessors had been noted. In reaction against this there were attempts to reform Benedictine monasticism and restore its observances. Most of these reforms gave rise to new monastic orders, of which the Cistercians were the most efficient at disseminating technical information.

The Cistercian Order originated at Citeaux in Burgundy in 1098, where, in an uninhabited, uncultivated area of forest, a group of monks settled with the aim of following the Rule of St. Benedict more strictly than was possible in the prosperous monastery from which they had come. From Benedict's instructions they learned that they should live separately from other people and devote part of their daily routine to manual labor. So they felt that their monastery should not possess lands or wealth outside its own boundaries, but should be self-supporting in its material needs—and that their own labor should be the means of achieving this.

The little abbey at Citeaux remained an obscure foundation until the year 1112, when a young nobleman and some of his friends decided they had a vocation to become monks there. This man, known to us as St. Bernard of Clairvaux, provided the dynamism that, by the time of his death in 1153, had led to the foundation of 340 Cistercian abbeys all over Europe.

In a Cistercian monastery, there were particular times laid down during each day for different activities. The lay brothers were supposed to devote seven or eight hours to manual labor, and the fathers four or five hours; this did not differ much from the Rule of St. Benedict, which allocated five and a half hours to manual work.[6] Intended as part of a spiritual discipline, the work done was supposed only to make each abbey self-sufficient. But it was often done so effectively that products like wool and iron were turned out in far greater quantities than were needed by the monks themselves. There was a market demand for these things, although the monks were warned, "Let not the sin of avarice creep in; but let the goods always be sold a little cheaper than they are sold by people of the world, that in all things, God may be glorified."[7] This injunction was perhaps too finely balanced, for the abbeys were soon in receipt of a large money income. In the course of time they became very wealthy, and the monks found themselves acting as estate managers and businessmen rather than laborers.

But that was much later, and what concerns us now is the way in which technical information could be passed from one abbey to another as part of the normal system of communication within the Cistercian Order. Fountains Abbey in Yorkshire may be cited as an example, for when it was founded by a group of Benedictine monks who had decided to adopt the Cistercian rule, they sent messages to St. Bernard, asking his advice. As a result, a certain Geoffrey came from Clairvaux to instruct them in the new rule. When money became available for building on a large scale, it seems to have been Geoffrey of Clairvaux who designed the new abbey, instructing the monks about the plans of Cistercian monasteries elsewhere, which were very standardized, owing partly to the strictness of the rule. The rule, for example, forbade the building of towers, though Fountains and Kirkstall Abbeys later acquired them. Churches were not to be of excessive height either, and should not have porches, sculpture, or stained glass. But the actual construction of the buildings was a technical matter in which the monks would be free to use the best methods available locally, so long as these did not involve excessive decoration. Thus at Fountains there is evidence that English masons erected the buildings, using their own methods for most detailed work, while probably only the overall design was by Geoffrey. But his influence serves to illustrate the speed at which technical ideas were transmitted within the Cistercian Order, for some buildings at Fountains completed in 1147 were tunnel-vaulted in the same way as comparable buildings put up at Clairvaux in 1140.

Pointed arches were used by the Cistercians from the 1120s onward, and the pointed tunnel vaults at Clairvaux and Fountains became very characteristic of Cistercian architecture at this time, although technically these vaults were not so advanced as the rib vaults first conceived at Durham in 1093.

Building is one technological activity of the Cistercians for which good evidence remains, but they also had an interest in other technologies. Kirkstall Abbey near Leeds was famous for the forge started there about the year 1200, and elsewhere in England the Cistercians played an important part in the development of iron-making. Iron was needed for agricultural implements and by soldiers for armor and weapons, but it appears that the Norman Conquest of England had largely disrupted the activities of existing smiths. The Cistercian monks in England played an important part in reestablishing furnaces for smelting iron ore. They seem to have prospected for ore and certainly exploited the main ironstones in northern England fairly systematically. The monks at Fountains worked both iron and lead in Nidderdale, and Furness Abbey exploited the hematite ores of the west coast, becoming famous from 1235 for the high quality of iron produced there. Most Cistercian abbeys had some mineral rights, and in the Forest of Dean, Tintern Abbey is known to have acquired a forge soon after its foundation in 1131.

Methods of smelting iron were of course primitive. The iron ore, charcoal, and limestone were heated together. Iron was then taken from the furnaces in the form of a soft, but not molten, mass with a great deal of slag mixed in it. By repeated hammering the slag could be forced out, and the iron was shaped into bars and plates suitable for the blacksmith to use for making tools or weapons.

This process was partly mechanized by using the same principle as the fulling mill, where waterwheels operated hammers by means of a series of cams or trips fitted to the waterwheel axle. A similar mechanism could be used to cause a hammer repeatedly to strike an anvil. Hammer forges of this kind were installed by Cistercian monks in both England and France. Indeed, it may be that one of the first forges with a water-powered trip hammer was at Fontenay Abbey. It is certainly significant that nearly all the information we have about hammer forges operating before 1200 comes from documents written by Cistercian monks.

However, agriculture and animal husbandry were always the most important forms of productive work with which the Cistercians had to be concerned. Since they normally established monasteries in remote and wild landscapes, they were responsible for bringing vast areas of virgin land into economic use. At first, in the twelfth century, their pioneering work benefited the whole economy. But as time went on, their concentration on sheep farming led to a very unbalanced pattern of agricultural development. The three largest Cistercian abbeys in northern England between them owned 45,000 sheep in the thirteenth century. The immense tracts of land reserved for grazing these flocks prevented the expansion of arable farming, and so contributed to the problems of inadequate food production that were becoming apparent here and in other parts of Europe around 1300. Another restriction that gradually became more important was that the monks sometimes retained a monopoly on the use of grain mills.

Water supply was another technique in which the monks were skilled, but here they made things easy for themselves by siting their abbeys near rivers. Then, by means of a series of conduits and canals, water could be diverted from the river to supply fish ponds, the abbey kitchens, and the *lavatorium* where the monks washed. After passing through these places where fresh water was required, the water from the river would flow beneath the *rere-dorter* or latrines, continually flushing them clean. Usually the water was made to flow from one building to another along stone-lined channels and drains, but lead pipes would sometimes be used. An example of the latter, associated with another order of monks (not Cistercians) is the twelfth-century water system installed in buildings attached to Canterbury Cathedral.

In Cistercian abbeys, the planning of water-supply channels also had to take account of waterwheels driving mills and forge hammers. The abbeys were often short of labor, and water power played an essential part in their economy. At Clairvaux, for example, the river drove corn mills, worked a fulling mill, a tannery, and various crushing mills, and supplied the kitchen and washing places before finally flushing clean the latrines (figure 4). At Fontenay, another French Cistercian abbey, the buildings that housed water-powered workshops still survive, with remains of the forge mentioned above. The best remaining corn mill in England is at Fountains Abbey, where a twelfth-century stone building measuring about 24 feet by 100 feet (7 by 30 meters) originally housed two waterwheels.[8]

The use of water mills was developed by other sectors of society than just the monasteries, and in the region of France to the north and west of Clairvaux, a total of 200 mills were in use during the thirteenth century. There had been 60 a century earlier, and only 14 in the eleventh century, so even allowing for gaps in the records of earlier periods, this represents a rapid increase.[9] Of course, many of these would be corn mills of the simplest kind, but a similar growth in numbers is revealed by a count of fulling mills mentioned in English documents.[10]

In addition to this extensive use of water power, the windmill (of the post-mill type) was introduced about 1150 or 1170; and the invention of the horse collar some 150 years earlier had allowed horses as well as oxen to be used as draught animals. So it is clear that one major characteristic of European technology prior to 1300 was a development of the use of power. Before that date, every possible source of natural power—rivers, tides, winds, animals—had been exploited. The amount of power available to supplement the unaided physical labor of men must

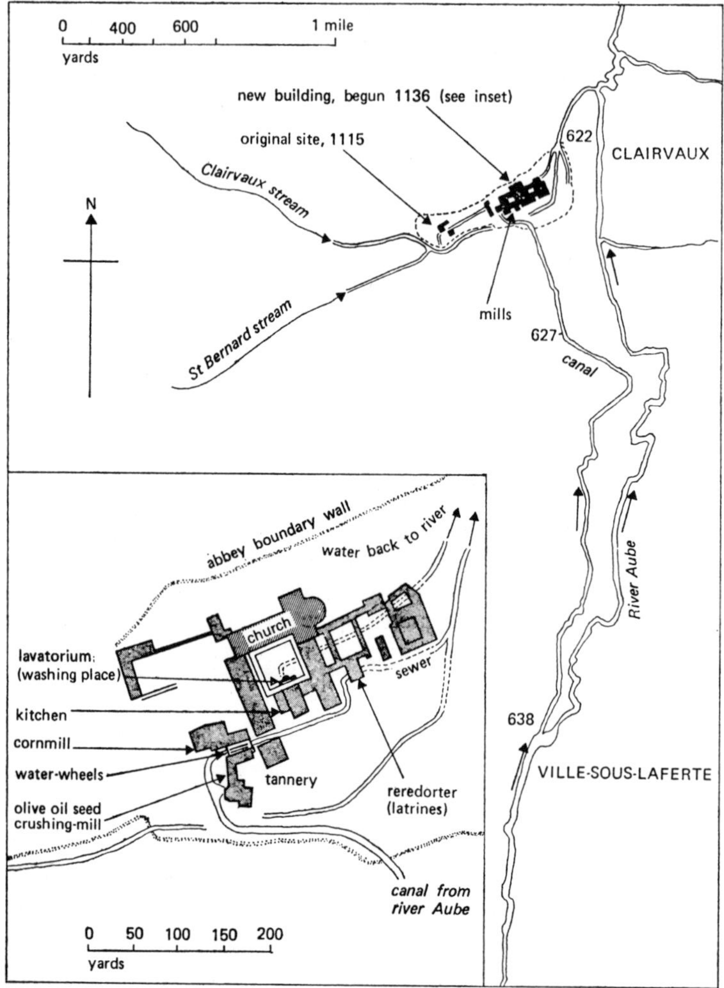

Figure 4
Clairvaux Abbey: a plan showing the water-supply system at its fullest development. New buildings, begun 1136, were sited about a quarter-mile (400 meters) to the east of the original site so that water from the River Aube could be used to supplement the flow from the two small streams.

therefore have increased by a large factor. This meant that a much higher level of material culture was possible than had been conceivable before, amounting to a change as momentous as the inauguration of the iron age. Other parts of the world, such as China and the Middle East, had gone through this same revolution in use of power two or three centuries earlier, but no other civilization had carried it through so intensively. Thus by 1250 or 1300, foundations had already been laid for the later technological ascendancy of Europe. To build on that foundation, much still had to be done to improve the very primitive machines of this period—but no further source of power existed for exploitation until the steam engine was invented.

Commerce and the Cathedrals

Work done in the monasteries was part of a spiritual discipline and was not aimed at producing goods for market. The improvements made by the monks in water supply, building techniques, and agriculture might therefore be seen as to some degree the result of an "idealistic" approach to technology. Though, as mentioned earlier, the monks did in time become businessmen and sell much of their produce, initially the direct influence of the profit motive on their practical and technical activities was probably very limited.

Not all of twelfth-century life was like this, however. There were commercial towns in northern France and Flanders where cloth manufacture was a major industry; and in northern Italy silk and cotton industries flourished, and merchants made fortunes in trade with the East. Ships from Venice and Genoa sailed to the eastern Mediterranean to bring back ginger, mustard, almonds, silk, and various metals. And the merchants who promoted these voyages set up joint stock companies and insurance schemes to protect themselves against loss. An example illustrating how joint stock companies began to evolve would be an individual merchant who, instead of putting all his money into one ship, would have, perhaps, a quarter share in four ships sailing on different routes. From as early as the 1070s, merchants were using this arrangement to ensure that even if one ship was wrecked, none of them was ruined, because other vessels would bring home a cargo and yield a profit.

In these circumstances, improvements made in shipbuilding and silk manufactures must have been strongly influenced by economic pressures. Thus innovation would be the product of quite different conditions from those in which monastic craftsmen worked; where commercial motives had probably been minimal in the earlier Cistercian abbeys, they would tend to predominate in the north Italian merchant cities.

The same would be true in the cloth trade of northern Europe. Raw wool bought by merchants from the English was spun and woven at Ypres and Ghent in Flanders, or in French towns such as Cambrai or Amiens. A typical weaver in one of these towns had perhaps a couple of looms in his house or workshop, which would be worked with the aid of one or two other men and an apprentice. In Ghent there were about 4,000 weavers, and since the total population cannot have exceeded 30,000, most of the labor force must have been employed in this trade. Flemish cloth was in demand all over Europe, and local merchants sent it to England or to northeastern Germany, from where their ships would return with corn, or with new supplies of raw wool. Trade over longer distances was organized mainly by Italian merchants, who bought cloth in Flanders, had it dyed and finished in the textile towns of northern Italy, and sold it in the eastern Mediterranean area.

It would seem, then, that in France and Flanders, as in Italy, there was a large area of life dominated by commercial activities. Any technical innovation, such as the introduction of an improved loom, would thus be influenced by the merchants' desire for larger output and bigger profits, and perhaps resisted by the clothworkers' need to protect their livelihood.

But however much the inhabitants of the cloth towns were preoccupied with the demands of commercial life, they did have idealistic and imaginative impulses that were capable of being mobilized with technological effect. And it was the Church that mobilized them with the most conspicuous success, as the cathedrals of northern France will show. Many of these great churches were built in the cloth towns and reflect the enthusiasm of merchants and artisans for enormous projects that the Church had set in motion.

Thus, at the risk of oversimplification, we can say that during the twelfth and early thirteenth centuries there was some technical progress motivated principally by economic considerations, and some that principally served idealistic purposes. Human motives are often very mixed, and to draw too sharp a distinction would be false. But contrasts can be drawn between the predominantly economic objectives of merchants in pursuit of their business and the predominantly non-economic aims of the cathedral builders, or, in their early years, of Cistercian monks. It is this kind of contrast that provides the main theme of this book, the argument being that idealism can provide a driving force for technical change—idealism embracing social or religious purposes, or even abstract symbols and intellectual schemes.

The Cathedral Crusade

The circumstances in which huge cathedrals were built in the commercial towns of northern France arose from the growing secular culture of those towns and the way in which the Church reacted to it. No longer were monks and clergy the only educated people to be found. Bankers, lawyers, and merchants had all, of necessity, to be both literate and numerate. And the universities founded at Paris (around 1150), Oxford (1190), Bologna (1200), and elsewhere were training men for the secular professions of law and medicine.

In 1096, the Church had effectively mobilized the idealism and the spirit of adventure in a restless and increasingly secular society by means of the first crusade. This military expedition to Palestine fired the imaginations of nobles and knights, to whom it was presented in the language of chivalry and religion, and for a while it absorbed their energies and resources in the service of the Church. The crusading ideal remained alive for two centuries and more, but subsequent crusades in 1147, 1187, and 1204 were less successful than the first and did not, in any case, appeal to the townsmen so much as to the nobility.

So a new kind of movement was needed in which the townsmen could participate, and such a movement was launched in the middle of the twelfth century. It happened about the time when the king of France was absent on the second crusade, and when the regent ruling in his stead was Suger, abbot of the great monastery of St. Denis near Paris.

Suger began to reconstruct the church building at St. Denis in the late 1130s, acting to some degree as his own architect. Every part of the structure he planned had some symbolic intention, as his writings show, and every element in its architecture expressed something of his philosophical or political ideals. As building work progressed, Suger was caught up in advance preparations for the second crusade, and the rebuilding of the church came itself to symbolize some of the ideals of a crusade. In the course of time, Suger's enthusiasm for building caught the imagination of other people. Between 1150 and 1280, about eighty cathedrals were built or rebuilt in France, many in the cloth towns in the north of the country. In effect, Suger had inaugurated a new sort of crusade, and one that was enthusiastically pursued by merchants and townsmen in the way in which the military crusades had previously engaged the enthusiasm of knights and nobles. Thus one historian has thought it right to talk about this wave of church building as "the cathedral crusade."[11]

Here, then was an idealistic movement with a constructive, material objective. What were its technical consequences? Briefly, they consisted in the development of techniques for building very high vaults supported by flying buttresses. The period of most active innovation was between 1160 and 1180, although many advances were made both before and after those dates.

To put the matter in a fuller perspective, it must be explained that the structure of the abbey church of St. Denis made use of some of the innovations that had first appeared at Durham, particularly the pointed arch and the rib vault. However, the architectural scheme to which these elements contributed was so novel that the rebuilding of St. Denis may justly be regarded as the beginning of the new system of architectural design that is now known as Gothic.

In this respect, though, its novelty was shared to some extent by Sens Cathedral in Burgundy, where a rebuilding scheme was begun in the 1130s. Work was not finished at Sens until 1160, while at St. Denis it was virtually complete in 1144. The common features of the two buildings owe much to the personal friendship that existed between Abbot Suger and Bishop Henry of Sens. The two men clearly discussed architectural matters, although in some respects Sens developed along different lines from St. Denis. Its design, in fact, had an important formative influence on architectural development. William of Sens came from there to England in 1174 to rebuild Canterbury Cathedral, and English architecture gained much from his example.

It is worth comparing the churches at St. Denis and Sens with the cathedral at Durham, because the two later buildings contain the first major development of the rib vault since the time, about forty years earlier, when it was introduced at Durham. In both of them, the pointed arches and rib vaults were used to create a structure with a far lighter and less massive appearance. For example, where the main supporting piers at Durham had been made in the form of heavy drums of stone, at Sens the corresponding members consisted of slender columns paired together. The resulting sense of lightness in structure was the essence of Gothic. The new style and technique of building was tried out tentatively in a few more French churches, including particularly Noyon Cathedral (begun c. 1150), but it began to show its full potential only in the 1160s, most notably in the building of the cathedral of Notre Dame in Paris.

At Notre Dame, one can see two of the most spectacular features of construction that were characteristic of the cathedral crusade—the systematic exploitation of flying buttresses and the tendency toward enormous size, especially in the height of the vault. Whereas, prior to the 1160s, the vault of a large church might be 70 or 80 feet (about 23 meters) above the floor as at Sens, at Notre Dame it was 110 feet (34 meters). The church was made exceptionally wide as well, with an extra aisle on each side.

Since its introduction at Durham, the flying buttress had not developed very far, because there were very few buildings in which its use was really necessary. But the extra height of the vault at Notre Dame increased the need for some kind of buttressing that would counteract the outward thrust of the vault. At the same time, the desire to refine the solid supporting members *within* the building meant that the use of an elaborate system of flying buttresses was essential *outside*. Thus after the start of construction at Notre Dame in 1163 and up to about 1180, there must have been a greal deal of intensive work done on the design of what was effectively a new system of buttressing. The aim was to reduce the weight that had to be carried by the internal walls of a church so that very large windows would be possible.

During the next seventy years, flying buttresses became extremely elaborate and sophisticated structural devices, and the characteristic appearance of a French cathedral when seen from the east was that of a forest of buttresses and their supporting piers, which almost entirely obscured the windows and walls of the building itself. There might be flying buttresses at two or even three levels, each aligned to serve a specific function.

At the highest of the levels, the buttress would act as a brace to the roof, strengthening it against wind pressures. Lower down, the main buttress would be arranged to support the vault.

All this is characteristic of the most advanced phase of medieval construction, which was reached in the mid-thirteenth century, after the completion of Notre Dame and of the cathedrals at Chartres (begun in 1194), Rouen (1200), and Rheims (1211). These are often considered to be the best examples of the High Gothic style and among the finest of all the medieval cathedrals, but developments in structural techniques that followed their completion led to an even more spectacular kind of architecture. The aim was now to create a church inside which no solid masonry walls could be seen at all, but only slender shafts soaring upwards with immense areas of stained glass in between; and for a ceiling, there would be an array of stone ribs echoing the shafts that rose up to support them from below. Only outside such a building would one be aware of the immensity and complication of the buttresses needed to support it. Apart from the use of flying buttresses, this architecture was made possible by the invention of window tracery, a part-technical, part-aesthetic innovation that allowed really big windows to be made.

The strong vertical emphasis of such an interior could be strengthened by adjusting its proportions so that it would appear high in relation to its width; but as figure 5 shows, the actual height inside cathedrals was progressively increased, and height itself seems to have become a matter for competition between one architect and another, and between one town and another. Thus, while Chartres Cathedral was of about the same height internally as Notre Dame (110 feet; 33 meters), the vaults at Rouen and Rheims were about 125 feet above the floor in both cases, while Amiens Cathedral, begun in 1220, had a vault soaring up to 140 feet above the floor. There was rivalry between the citizens of Amiens and the men of Beauvais, two of the bigger commercial towns in northern France, so when the rebuilding of the Beauvais Cathedral was begun, in 1247, it was felt to be essential that the dimensions of Amiens Cathedral should be exceeded. The vaults at Beauvais were therefore pushed up to the extraordinary height of 157 feet (48 meters). The dimensions of Beauvais were equalled by Cologne Cathedral, which, despite its distance from northern France, is a closely related design. Like Westminster Abbey (rebuilding scheme begun 1245), Cologne Cathedral (1248) was partly the

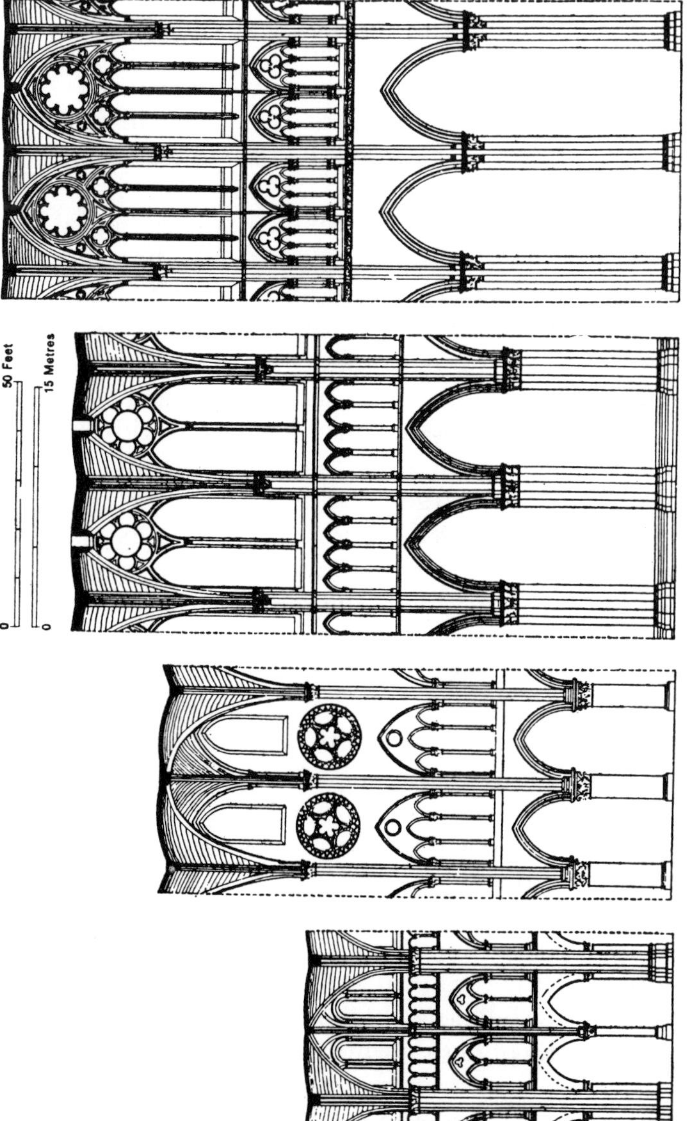

Figure 5
Bays from four French cathedrals, drawn to the same scale and showing the increasing height of vaults. From left to right, Noyon Cathedral, designed c.1150; Notre Dame, Paris, as designed c.1170; Rheims Cathedral c.1211; and Amiens Cathedral, begun 1220. (Source: Figures 66, 69, 76, and 77 from *An Outline of Euorpean Architecture*, by Nikolaus Pevsner, Penguin Books, 7th edition, 1963, copyright © Nikolaus Pevsner, 1943, pp. 98, 100, 106, and 107. Reproduced by permission of Penguin Books Ltd.)

result of French architects working on foreign soil; but while Westminster Abbey is a modest, anglicized version of Rheims, at Cologne the interest in enormously high vaults, vast expanses of glass, and soaring vertical shafts had its most sensational result.

The same interest in building to great heights was expressed on the exterior of the buildings by towers surmounted by spires. The cathedral at Chartres, for example, now impressive enough with its two spires, was originally intended to have eight of them: likewise Rheims was to have six spires, while at Laon, five towers were completed. This "vehement verticalism" of the exteriors, "the supreme expression of the heavenward urge,"[12] was in most cases still to be realized when the cathedral crusade exhausted itself in the 1250s and 1260s.

After this time, there were no developments in cathedral architecture with any wide technological significance. The structural sophistication of Amiens, Beauvais, and Cologne Cathedrals was the culmination of a long series of experiments in which many different structural techniques were tried out. Those that were successful were used again and again and improved each time, while those that were associated with the subsidence or collapse of part of a building were abandoned. Improvement must have been largely a trial-and-error process, but what is noteworthy is the continuing willingness to experiment. A similar courage and openmindedness, and a similar spirit of inquiry, must have pervaded other technical arts at this time and must account for the improvements and inventions made in them, but only in cathedral building can it be so clearly observed.

All this raises questions about the extent of a thirteenth-century mason's understanding of the engineering problems he faced. Obviously he lacked the kind of knowledge a modern engineer would have, yet it is clear that the masons usually knew very precisely where buttresses ought to be and at what height their counterpressures were needed. To this extent, then, their experiments were guided by knowledge. Their trial and error was not just groping in the dark, for they had a considerable insight into what shape a building should be: insight that was aided by geometrical rules of design which will be mentioned in chapter 2. And as it turns out, the successful design of masonry structures depended more on getting the shape—that is, the geometry—right than on calculating the magnitude of forces in the modern way.

Now of course, cathedral building is not technology in any normal modern sense; it is an art form, expressing religious truths through symbolism, and religious emotion through mass and space and soaring height. It expresses other things as well—civic pride and mercantile prosperity—but also a mood of restlessness, of never being content with what has been achieved, whether in the moral, or the aesthetic or the material, spheres. It gives evidence too of an enthusiasm for experiment in both aesthetic and structural matters, as if the builders, always unsatisfied by their handiwork, were continually lured on by higher ambition. There was almost a moral imperative about it, rather of the sort expressed by St. Bernard in the twelfth century when he said that "a man who sets before himself a greater good and then does a lesser, sins."[13] The cathedral builders had set before themselves a very great good, but never accepting that they had attained it, they were continually driven on to explore and experiment, and to attempt the untried.

They were spurred on too by the competitive spirit that arose among the independent commercial cities, and since the citizens of these towns clearly made comparisons between their own and neighboring cathedrals, the progress made in terms of size or aesthetic effect must have been quickly appreciated. Thus although it cannot be claimed that the townsmen were conscious of technical progress as distinct from changes in architectural style and aesthetics, we can say that they were deliberately and successfully promoting change in an art that had a strong component of engineering and mechanical skill. Thus, possibly for the first time in history, a society was consciously committed to systematic and deliberate change in a wide range of practical arts. In this restricted sense, the cathedral crusade can be taken as the beginning of the modern phase in the history of technology—the phase in which the development of techniques moved from the evolutionary stage: that is, from a state in which craftsmen make occasional small changes in their art, and from one generation to another gradually improve their skill and the stock of inherited experience, to a stage where changes are deliberately made in order to approach some unrealized ideal that is always one step beyond what current techniques make possible. The continually questing mood of the cathedral builders, at any rate, is characteristic of the spirit that brought modern technology to birth.

The Cost of the Cathedrals

The enthusiasm of medieval society for their cathedrals can be assessed to some degree by the money spent on building them. Some of the best information for this it to be found in the accounts of the Office of the King's Works in England. But in studying these accounts it is as well to remember that thirteenth-century prices have to be multiplied by something more than a thousand before comparisons with modern values can be made. Westminster Abbey, for example, where a rebuilding scheme was begun in 1245, had cost about £41,250 by 1272, but only about two-thirds of the building had been completed for that sum. Had it been completed in the thirteenth century, its total cost would perhaps have been £70,000. This figure can be put into some sort of perspective by comparison with the series of great castles begun in north Wales in 1283. Of these, the one at Beaumaris had by 1300 cost £11,400 and needed a further £3,000 for its completion, while Conway Castle and town walls were finished for a total cost of about £15,000. It is interesting to note that the man responsible for supervising the construction of the Conway walls was one of the rather few people of this period who called themselves "engineers," referring especially to military engineering.[14] Known as Richard L'enginour (fl. 1277–1315), this man also built bridges and pontoons for river crossings in north Wales and constructed siege engines. Later he settled in Chester, where he managed a group of water mills.

It would seem from this that at the end of the thirteenth century, £15,000 in English currency would build a large castle, whereas four or five times as much was needed for a church of cathedral size. Part of the difference would be accounted for by the sculpture and stained glass and perhaps the shrine of a saint. Nonetheless, much of the great cost of a cathedral was due to the refinement and sophistication of the structure. Stones used in the construction of arches, buttresses, and vaults needed to be very accurately cut, and so were more expensive than the much larger quantities of stone used to build a castle. The cost of a cathedral was therefore a reflection of the great concentration of specialized skills involved.

The question arises as to how this money was raised during the cathedral crusade. For although the French king made gifts to some cathedrals, such as Chartres, and in England, Henry III paid for Westminster Abbey, contributions of this kind could not always be expected, and it was within the towns where the cathedrals were built that much of the money was raised. Local merchants would probably be the biggest contributors, and they would usually be granted indulgencies in exchange for their benefactions. In addition, the guilds or trade corporations in a town would each donate a window or some other specific item for the cathedral. But it was by the exhibition of saints' relics that the great mass of the population was reached, and this brought in money from even the poorest people, as well as from pilgrims who traveled great distances to see such things.

In 1124, when Abbot Suger began raising funds for the rebuilding of his abbey, St. Denis, near Paris, he estimated that three-quarters of the money would be contributed by pilgrims, and the ordinary revenues of the abbey would make up the rest. There was inevitably a minority of people who questioned the spending of money on this scale for such unnecessarily large constructions, and they could quote St. Bernard, abbot of the Cistercian monastery of Clairvaux, who criticized monks of other orders for "the vast height of your churches, their immoderate length, their superfluous breadth, the costly polishings, the curious carvings and paintings which attract worshipper's gaze." Bernard went on to criticize the way in which the money for all this was raised by the exhibition of relics. People's eyes, he said, "are feasted with relics cased in gold, and their purse strings are loosed. They are shown a most comely image of some saint, whom they think all the more saintly that he is the more gaudily painted. . . . The church is resplendent in her walls, beggarly in her poor; she clothes her stones with gold, and leaves her sons naked."[15]

St. Bernard wrote a great deal in this vein and did his best to ensure that Cistercian abbeys had plain, undecorated buildings, limited in size to what was strictly necessary. There was, indeed, a strong streak of puritanism in Bernard's attitude, reflected also in his more strictly religious writings. And the similarities of Bernard's view with those of the later Puritans who were products of the Reformation is so striking that one historian has used the phrase "the high ancestry of puritanism"[16] to describe the Cistercians. One reason given for the similarities is that both

Cistercian monks and later Puritans were reformers of the kind who wished to revive the customs of the early Christians. St. Bernard spoke of the church as it had been in "days of old, when the apostles spread their nets to take not gold or silver but the souls of men."[17]

From this desire to revive the spirit of the early Church, then, sprang the injunctions about plain and simple buildings, and an austere way of life. Connected with it was also the idea that work—manual labor— had a moral value and was part of one's vocation. In all this, the comparisons that can be drawn between the ideals of twelfth-century monks and seventeenth-century Puritans are worth mentioning because of the link between a high valuation of work and a progressive outlook on technical improvement that can be observed among both groups. Of course, there were important differences as well, but the similarities are sufficiently close to give pause to those historians and sociologists who identify the work ethic solely with the later Puritans and who refer to it as a "Protestant ethic." One other important point is the way in which economic pressure on technical progress was relatively limited in extent during the twelfth century, so that some kinds of innovation came about largely as a result of ideological influence. We have discussed *two* particular groups of people for whom this was partly true—Cistercian monks and the cathedral builders—and there were two prominent men who were outstanding for their exposition of the ideals of the two groups—St. Bernard among the Cistercians and Abbot Suger at the outset of the cathedral crusade. The two men knew each other personally and discussed their differing ideals. In their attitudes to the crusades and in politics generally, they represent two aspects of the idealistic trend in practical and technical work that prevailed during the twelfth century.

The success of fund-raising in support of the cathedral crusade meant that masons and builders could pursue this trend in erecting ideal structures without being too tightly bound by economic constraints. Improvement of the technical means of erecting a cathedral could thus become an end in itself, justified only by the symbolism inherent in any church building. In this context, inventions were neither motivated by economic necessity nor selected for development with profit in mind, but were largely a response to the ideological purposes of the cathedral crusade.

Beyond the world of cathedrals and abbeys, there was, of course, the life of commerce—the merchant ships in the Mediterranean and the North Sea, the clothworkers of Flanders and the silk workers of Italy. There was also significant technical innovation in these areas of economic life, some of which will be discussed in the next chapter. There is a striking contrast, though, between economic objectives in technical progress and the influence of ideas and idealism. The different kinds of objective—economic or idealistic—represent different directions which technical progress could take. "Progress" indeed is a misleading word to use about technology, because it implies steady advance toward a single, universally accepted goal. In fact, the direction of technical progress has varied in the past. The choice of direction is always open, not least in the twentieth century, and it is important to appreciate the full range of options that are available.

The different directions that technical progress took during the period covered by this chapter may be illustrated, somewhat crudely, by means of a diagram (figure 6). Progress in some of these directions led to innovation that was "useful" in economic terms. But sometimes, as in the monasteries, the "usefulness" of an invention was, in theory, a matter of spiritual criteria, while among the cathedral builders, technical progress took a direction which had no practical purpose at all, and was of no "use" to anyone. The technical effort that went into the cathedral nonetheless had immense symbolic importance.

In this chapter, a rather general view has been taken of these different kinds of progress. But it is also necessary to study them in a more personal and individual way and to ask, What was it that stirred in the imagination of people at this time? That may be unanswerable, but it is a question worth considering, because invention depends on the exercise of imagination. Idealism is important in the history of technology partly because ideals are often a more effective trigger to the imagination than are economic incentives. The next chapter will look at a number of especially imaginative aspects of the technology of this period in the belief that they can provide clues as to what it was that kindled the restless, creative spirit out of which modern technology was eventually brought to birth.

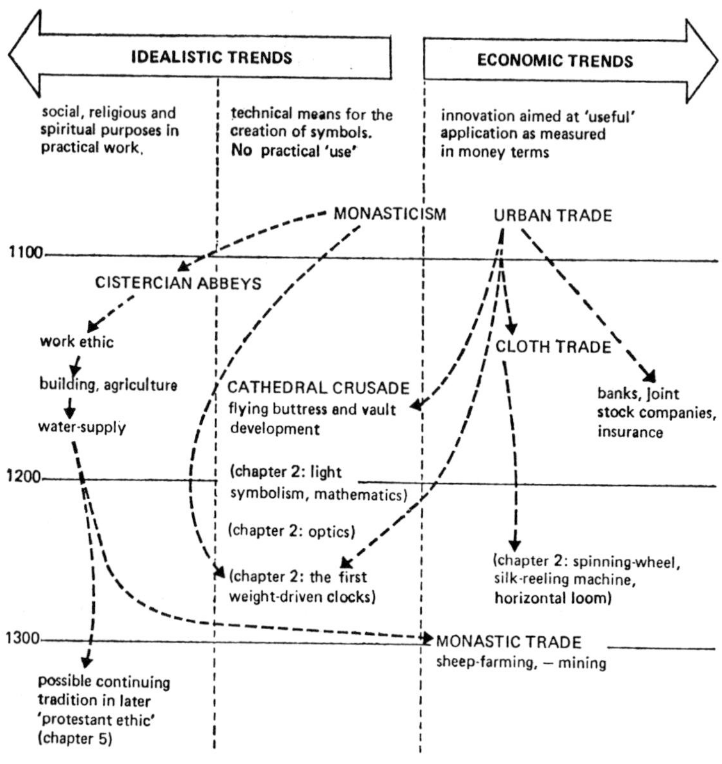

Figure 6
Schematic representation of different directions taken by technical progress during the twelfth and thirteenth centuries.

2

A Century of Invention: 1250–1350

Prospects for Progress?

It is often important in studying the history of technology to know what people believed about "progress" and whether they saw it as related in any way to the development of technology. During much of the history of Christian civilization in Europe, the modern idea of progress hardly existed, but instead many people had a view of human destiny based on the Bible and in particular on its final book, the Revelation (or Apocalypse) of St. John the Divine. Part of this was devoted to what seemed a prophecy concerning the end of the world. If this was soon to happen, what could anybody believe about progress? And what was the significance of the new techniques people saw being developed by cathedral builders, by textile workers, and others, if there was no long-term future for civilization?

One thirteenth-century writer who believed that the end of the world would not be long delayed, but yet had a strong interest in current inventions, was the Franciscan friar Roger Bacon. In many of his writings, Bacon gives a strong impression of advocating technical progress. But in common with most of his contemporaries, he took a pessimistic view of humanity, among whom, he thought, corruption and evil would increase until all was swept away at the end of the world. The forces of evil, including the Antichrist himself, would use all manner of natural and artificial devices to confound mankind, and it was necessary for men to improve the practical arts in order that this threat be overcome. Similarly, the extension of knowledge was necessary as a means "of preparing Christendom to resist more successfully" the corruption that was coming,[1] which would precede the end of the world. The technical progress and the search for knowledge that Roger Bacon advocated had a special purpose and was strictly limited in time by the forthcoming apocalypse. So although he expressed an enthusiastic interest in the technical achievements of his age, he could not state any belief in a sustained technical progress.

The idea of progress was "glaringly incongruous" with Bacon's understanding of the world. "When he looked forward into the future, the vision which confronted him was a scene of corruption, tyranny and struggle under the reign of a barbarous enemy of Christendom; and after that the end of the world. It is from this point of view that we must appreciate the observations which he made on the advancement of knowledge."[2]

The same attitudes were expressed by the cathedral builders in the sculpture that formed part of their work. Very frequently, and on prominent parts of a building such as the main doorway, they depicted the end of the world and the Last Judgment. It seems reasonable to suggest that their views were in some ways similar to Roger Bacon's, and that their efforts to improve technical skills were not conceived as a means of improving the condition of man within the present order of things; but rather, they were reaching forward to meet an eternal order, a New Jerusalem, which the cathedral itself symbolized. The conviction was that by "reaching out to the immaterial through the material, man may have fleeting visions of God."[3] This, perhaps, was the meaning of the technical progress Roger Bacon advocated, and which evoked the enthusiasm of the men who built—and paid for—the cathedrals.

How cosmic speculations and theological dogmas could affect the mundane activities of the artisan will be discussed in more detail toward the end of this chapter. For the moment it is sufficient to draw a contrast between the views of Roger Bacon and the attitudes current in the sixteenth century. By that time, technical progress was appreciated in a humanistic sense; but the evidence quoted to show that real progress had been made in western Europe most frequently took the form of lists of inventions, including, ironically, a large number of thirteenth- or early fourteenth-century date—the mariner's compass, gunpowder and firearms, clocks, and the Italian silk-twisting machines. Of the other inventions commonly mentioned, only printing and such specialized items as the Toledo waterworks belonged to the period when technical progress was coming to be recognized. Thus even if the men of the thirteenth or early fourteenth century did not believe in technical progress, that belief as it was first formulated in the sixteenth century was based very largely on their achievements.

Firearms and Textiles: Some Inventions of c. 1250–1350

The enthusiasm of the cathedral builders seems to have flagged soon after 1250; fewer large churches were built and the pace of innovation slowed down. Nonetheless, the period between 1250 and 1350 was one of prolific invention in some of the other technical arts. Here the historian faces a difficulty, however, in that none of these inventions can be documented as precisely as, for example, the earlier series of minor innovations that marked the development of the flying buttress. All that can be done in some cases is to say that most of the crucial stages in their development took place in the century between 1250 and 1350. Thus, for example, it is not known when the first weight-driven clock was successfully made, but it is clear that by 1350 several large astronomical clocks had been built.

Some of the innovations of the period have been attributed to new contacts with China. Mongol forces had conquered most of Central Asia and southern Russia and had created conditions in which information easily passed westward across Asia. For a while, Chinese officials worked in Russia and Chinese military engineers in Persia, while several Europeans visited Mongolia or China, including Joannes de Plano Carpini (1246), William of Rubruck (1253), and Marco Polo (1260).

Through these or other channels, Europeans learned about gunpowder, which the Chinese had been using for three centuries, and it seems probable that Europeans also learned about the primitive guns that were being tried out in China from before 1288. Certainly, Europeans were experimenting with very similar guns soon after 1300 and very quickly developed a more formidable weapon than the Chinese yet possessed, the cannon. The details of these innovations have been discussed elsewhere.[4] Here it is sufficient to note that guns did not change the nature of warfare in any decisive way until later, and if attention is confined to the period 1250–1350, it was perhaps new devices for spinning and weaving that made the biggest material difference to European life. The Ypres "book of trades," dating from around 1310, provides some of the most interesting evidence, for it includes illustrations of most of the processes involved in making cloth. From it, we can see that one of the most important innovations in the weaving trade was the horizontal loom, which may, like many other things, have come from the Islamic civilization, perhaps via Spain. But in 1310 it was still a fairly recent invention and offered several advantages over earlier looms in which the cloth was woven in a vertical position. With the horizontal loom, some operations were controlled by means of pedals, leaving the weaver's hands free to pass the shuttle backward and forward.

It is worth recalling that the cloth trade at Ypres and throughout the Low Countries was based on wool but that in Italy there were also silk and cotton manufacturers. One rather impressive machine used in the silk industry was a device for twisting silk thread. Machines of this sort are first heard of at Lucca about 1272, and then at Bologna. One example of about 1330 had bobbins rotating on 240 spindles, with two rows of horizontal reels above. This seems to have been a standard size, since another is reported with the same number of spindles at about the same date, while one very large machine had 480 spindles.

Bobbins were fitted to the spindles, and the silk was twisted as it was wound from the reels onto the bobbins, the twisting action being performed by the rapid rotation of an S-shaped flyer around each bobbin (figure 7, with detail A for the flyer). Much of the silk-processing technology of northern Italy had originated with silk workers from Sicily, where there had been a considerable industry in Islamic times, and it is uncertain whether this machine was an Islamic or north Italian invention.[5]

There are no illustrations of these machines until 1487 and then c.1600 (figure 7), so there is also uncertainty about the working and construction of the earliest examples. However, it seems probable that the basis was a circular frame, like the example illustrated, with a large wheel inside the frame driving the bobbins and flyers. The power driving this wheel would sometimes come from a waterwheel, sometimes from animals.

The introduction of the silk-twisting machine was probably related to the invention of the spinning wheel in its European form, which occurred at about the same time. Here it must be explained that silk did not have to be spun in the usual sense. Individual fibers are very long, and to make a silken thread it is only necessary to twist several fibers together. In contrast, wool and cotton have relatively short fibers, and to make a thread or yarn from wool means not only twisting the fibers together, but also drawing out the yarn while it is being twisted in order to

Figure 7
An Italian silk-reeling machine drawn by Vittoria Zonca around 1600. The spindles are made to rotate by a large wheel or drum that revolves inside the circular framework, which in turn may be driven by a water-wheel. This machine is probably quite similar to the ones used c.1300 from which it is developed.

(Source: Vittoria Zonca, *Nuovo teatro di machine et edificii,* 1607; see also A. G. Keller, *A Theatre of Machines,* London: Chapman and Hall, 1964.)

lock the short fibers together. From a very early date, this operation had been performed by means of a spindle and whorl, or drop spindle, which really meant that it was all done by hand with the aid of only very simple tools and a spindle on which the yarn could be wound.

The earliest spinning wheel in Europe, introduced during the thirteenth century, consisted simply of a frame to support the spindle and a wheel and pulley to make it revolve. The twisting and drawing operations in the spinning process were still a matter of the manual skill and judgment of the worker. This type of spinning wheel was known as the "great" wheel to distinguish it from a second type, the "Saxony" wheel, in which the simple spindle was replaced by a bobbin and flyer device. By this means, the twisting part of the process was mechanically controlled, and it became possible for the thread to be wound continuously onto the bobbin as it was spun.

The earliest reference to a great wheel dates from 1298, when these machines were in use for spinning the weft (but not the warp threads) of cloth made at Speier on the Rhine. On the other hand, the bobbin and flyer was not illustrated anywhere earlier than a German work of around 1480 and a little later in Leonardo's notebooks. Most authors take this to mean that the Saxony wheel, with its bobbin and flyer, was introduced about two centuries later than the great wheel. The matter is open to doubt, however, because Italian silk-twisting machines of the thirteenth century almost certainly employed the bobbin and flyer, and as Patterson[6] points out, early "Bolognese twisting mills are closely analogous to spinning devices," and these mills were in use before 1300. Similarity in working principle between machines used for wool and for silk has therefore led Patterson to wonder whether the spinning wheel, and particularly the Saxony wheel, may have been introduced considerably earlier than the sparse documentary evidence suggests. He has also pointed out that the earliest pictures of bobbin and flyer devices show a fully developed mechanism, so there must have been simpler and more experimental versions of the same thing long before these sketches were made in the late fifteenth century. For these reasons it seems fair to suggest that the Saxony wheel should be grouped with the great wheel and the Italian silk-twisting machine as one of the inventions of the period between 1250 and 1350.

The Weight-Driven Clock

Of all that was achieved in the practical arts during the period under review, the invention of the weight-driven clock has attracted most attention from historians. Part of the reason is that although documentary evidence is small in quantity and difficult to interpret, there is very much more known about the clock than about almost any other contemporary invention. But another reason for emphasizing it is the way in which this invention is supposed to have led to new concepts of time, making possible the whole pattern of modern life, in which so much is regulated by the clock. On another score, there is sometimes excited comment about the way in which the clock has on occasion been taken as a paradigm of all machinery, so that it is seen as a sort of parent to the whole of our mechanically minded civilization.[7] On a more technical level, it is rightly pointed out that the clocks were the first machines to be made entirely of metal; all other medieval machinery was made mainly of wood, whereas early clocks not only had metal gear wheels rather than wooden ones, but their mechanism was supported in an iron framework.

Those who argue that the weight-driven clock gave rise to new concepts of time and more precisely regulated habits of life usually point out that in the thirteenth century it was monks who most often needed to know the time. This was because of the offices which were said in church at regular intervals. The Rule of St. Benedict, compiled in the sixth century, gave instructions about the times of these offices and referred to them by names that indicate the method of counting the hours then in use. The interval between sunrise and sunset was divided into twelve hours, so that each hour during daylight was longer in summer than in winter. The office said in church at sunrise was called Prime, and other services were Terce (at the third hour of daylight), Sext (at the end of the sixth hour: that is, at noon), and None (at the ninth hour of daylight). There were also two services at about sunset, the times of which could easily be determined by watching the sky. Thus the problem of measuring time was most acute for the services said during the day, and also for one that was said in the eighth hour of the night. The usual time-measuring instruments in the twelfth century were sundials, hourglasses, and water clocks, and the monks developed water clocks to particularly good effect. The flow of water through a small hole provided the measure of time, and such devices, equipped with alarm mechanisms, were often used to wake the monks for the night office. There is a story that in 1198, at the abbey of Bury St. Edmunds, the clock woke the monks during the night just in time for them to put out a fire that had started in their buildings. What is more, the clock contained sufficient water for it to be of considerable help in extinguishing the flames.

This story suggests that a water clock could be fairly large, or at least could include a fairly large vessel of water. The alarm mechanism would be set in motion when the water in this vessel reached a particular level; then a weight would be released, whose fall would in turn ring a bell. In a machine of this kind, one of which is described in an eleventh-century manuscript, the weight was attached to a cord, and as it fell, unwinding the cord, it made a drum rotate. This probably worked a mechanism by which an iron bar was made to strike the bell repeatedly. It has been suggested that the origin of the weight-driven clock is to be found in a modification of this mechanism, where the bar that struck the bell evolved into the "foliot" bar of an escapement. Here it must be explained that the earliest weight-driven clocks were made to keep time, rather approximately, by a mechanism known as the "verge and foliot." This was a rather curious affair, in which a heavy bar (the foliot), pivoted near its center, was pushed first one way and then the other by the action of a toothed wheel, which was only free to move through the space of one tooth for each vibration of the bar; the wheel was driven through simple gearing by a weight suspended from a drum (see figure 8). Whether such a clock ran fast or slow depended largely on the size of the weight; other weights could be added at the ends of the foliot, and the size and position of these would give a further "fine" adjustment. The earliest *surviving* clocks in which the foliot mechanism was used are those belonging to the cathedrals at Salisbury (1386), Rouen (1389), and Wells (1392). The famous Dover Castle clock is probably rather later.

This, very briefly, is the way in which the invention of the weight-driven clock can be explained if only utilitarian needs for a timepiece are considered. But it seems doubtful whether questions of utility can really account for this invention, because in practice early weight-driven clocks were neither reliable nor accurate, and water clocks were probably more satisfactory for time-keeping. The first weight-driven clocks were more proba-

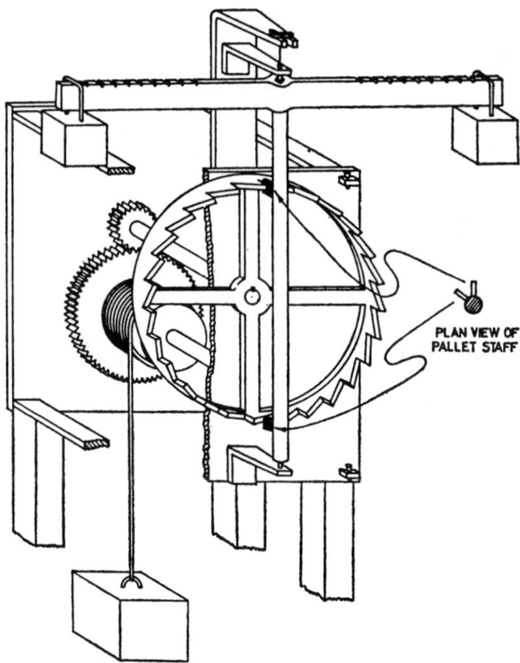

Figure 8
Principle of the "verge and foliot" mechanism. The horizontal iron bar with weights at either end is the foliot. Fixed to its center point, immediately below the thread suspension, is a vertical spindle, referred to on the diagram as the pallet staff. This is the verge. It carries two projecting pieces of metal, the pallets, which engage with the teeth of the crown-wheel in such a way that only one tooth may pass the pallet during one complete swing of the foliot.

(Source: Aubrey F. Burstall, *A History of Mechanical Engineering,* London: Faber and Faber, 1963, fig. 72, p. 130. Reproduced by courtesy of Faber and Faber Ltd.)

bly made for reasons of the intellect or imagination independently of any utilitarian or economic incentive. There was an interest in mechanical things for their own sake, which was already strongly developed in Europe by 1300, but more than that, the first clocks provide an excellent example of how *ideas* may influence the direction of technical progress.

Some clue as to the kinds of idea that influenced the invention of the clock is provided by the earliest examples of which there is a detailed description. For while the earliest *surviving* clocks are the relatively simple ones already mentioned, which date from around 1390, the earliest clocks of which full details are available were far more complex; they were built by Richard of Wallingford at St. Albans around 1330 and by Giovanni de' Dondi at Padua in 1364. The main purpose of the two latter machines was to indicate the position in the zodiac of the sun, moon, and planets. The Padua clock did this by means of seven dials, one corresponding to each known planet, one for the sun and one for the moon. Now, devices that represented the positions of the planets had existed in many forms for a long time before the invention of the clock. Sometimes they were intended only to illustrate the principles upon which the planetary motions were understood; but often they were designed as computing machines that could be used to predict the positions of various heavenly bodies. The simplest form of computing device was the astrolabe, which was as easy to work as a modern slide-rule, but which yielded information about only the fixed stars. By adding gears and additional dials, Arab astronomers devised astrolabes which also showed the motions of the sun and moon. This seems to have been done by about the year A.D. 1000, but the earliest geared astrolabe of this type to *survive* is dated 1221, and was made in Persia. It had eight gear wheels, with teeth shaped like equilateral triangles, and the mechanism was turned by hand. The technical significance of this kind of instrument was that trains of gears had to be designed for it in which the relative speed of rotation of various wheels needed to be precisely specified.

During the eleventh century, Islamic astronomers also invented the equatorium, an even more complicated device used to calculate the positions of the planets. This was being used in Spain by about 1025; and when the expanding Christian civilization of Europe began to take over the Islamic territories in Spain during the late eleventh and twelfth centuries, knowledge of the equatorium passed into Europe along with much other astronomical information.

"No Islamic examples of the equatorium have survived, but from this period onward, there appears to have been a long and active tradition of them . . . they were the basis for the mechanized astronomical models of Richard of Wallingford (and Giovanni de'Dondi)."[8] They were the basis, in other words, for the clocks designed by these men. The faces of these clocks were not marked with the hours or used for telling the time but were the dials familiar to users of the astrolabe and, more particularly, the equatorium. The only difference was that these dials were turned continuously by the fall of a weight. They represented the fulfilment of an ambition to make powered astronomical models keep pace with the moving planets. There was no utilitarian reason for doing this; it was just that the idea of a man-made machine synchronized with the heavens had immense symbolic value. Such a machine was really itself an artificial universe.

The argument, then, is that the clock evolved through the gradual development of astronomical machines of this kind.

A Wheel to Move with the Rotation of the Heavens

This is an interpretation with which not everybody will agree. But even those who prefer a simple account of the invention of the weight-driven clock, to the effect that it was the result of improvements to monastic alarm mechanisms, must at some point discuss clocks in terms of astronomy. In the absence of a clock, people tell the time by watching the passage of the sun and stars across the sky, and the purpose of a clock is simply to provide a convenient indication of where the sun is at any moment. In some ways, then, every clock is a working model of the sky.

The special significance this idea had around 1300 can be understood only if one knows something about the picture of the universe that was then current. It was a view inherited from the Greek civilization of the distant past and was based on the idea of a spherical earth, fixed and motionless, at the center of the universe. Around the earth in circular orbits moved the moon, the sun, and the planets. Finally, just beyond the orbit of Saturn, the most distant planet then known, there was the sphere to which all the stars were fixed; this was the outer shell of the universe, so far as men could see, and it too revolved about an axis that was marked, very nearly, by the pole star. Thus while the stars could be seen to move together across the night sky, the pole star alone was almost stationary.

The luminosity of the heavenly bodies, and their untiring motions, seemed to show that they were of a different nature from anything known on earth. For on the earth there is no fire that burns continuously like the sun without being consumed, and there is no motion that does not run down and stop: none, that is, except the tides, and they were rightly thought to be influenced by the moon. And for a clock to be a true working model of this wonderful planetary system that gave men their sense of time, it would have to work rather as the tides, being somehow driven by the continuous motions of the heavens and keeping in step with them. The clock, if it worked perfectly, would have to run with a perpetual motion.

How widespread such ideas were among the inventors of the clock is very much open to question, but it cannot be denied that they played some part. This can be illustrated by the writings of a group of people who were active in the 1260s and perhaps earlier, among whom Pierre de Maricourt, author of an essay on the magnetic compass, provides the clearest evidence. Written in 1269, Pierre's essay was sent as a letter from Italy, where he had gone with the army of Charles of Anjou. It was addressed to a soldier friend living near Pierre's home in northern France, and while it is possible that Pierre was himself a soldier, it is usually assumed that his job in the army was that of a craftsman—an armorer or a builder of siege engines.

Maricourt and Foucaucourt (Fontancourt), the villages from which Pierre and his friend had come, were only a short distance from the home of Villard de Honnecourt, the stonemason or architect (see map, figure 9). Pierre and Villard had some interests in common, so it would seem quite likely that they were acquainted with each other. Pierre's circle of friends probably included Roger Bacon (c. 1215 to c. 1292), and perhaps also a certain Robert, author of a treatise on the astrolabe. Both these Englishmen spent many years in France.

Between them, this group of scholars and craftsmen provides a valuable indication of the thinking of men with a practical turn of mind. In particular, all four of them were interested in water clocks and similar mechanisms, and also in perpetual motion.

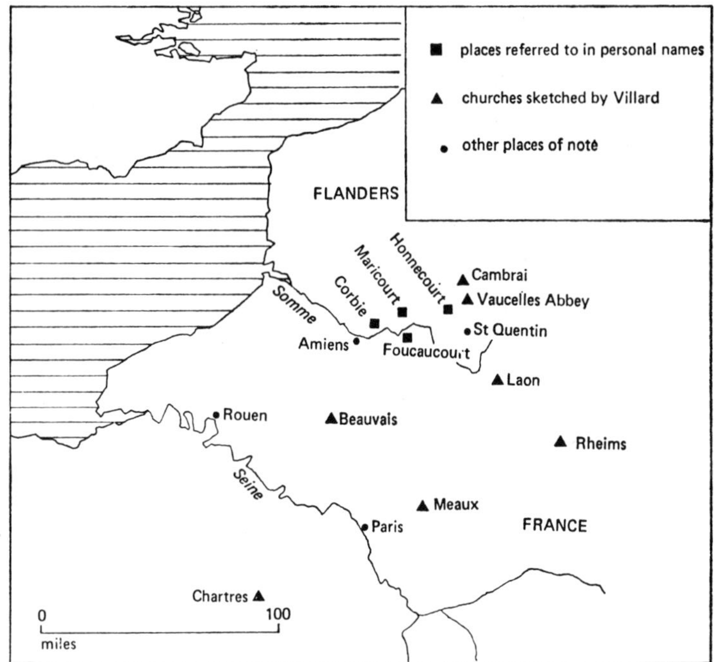

Figure 9
Places associated with Villard de Honnecourt and Pierre de Maricourt.

Pierre's comments on these devices form part of his letter on the magnet, in which he argued very cogently for the idea that the magnetic compass was linked to the heavens by some hidden force or "virtue." The magnet in a compass pointed always to the pole star, or almost so, and therefore must be linked with the axis of rotation of the heavens. The magnet had poles of its own, as Pierre showed, and ought therefore to have its own axis, about which it could turn with a perpetual motion.

To demonstrate this it was necessary to cut a magnetic loadstone into a spherical shape, pivot it at its poles, and align it with the axis of the heavens. "If now the stone be moved according to the motion of the heavens, you will be delighted in having discovered such a wonderful secret; but if not, ascribe the failure to your own lack of skill rather than to a defect in nature. . . . With such an instrument you will need no timepiece, for by it you can know the ascendant at any hour you please, as well as all other dispositions of the heavens which are sought for by astrologers."[9]

This device, then, was supposed to function as a clock, and it illustrates the connection between clocks, astronomy, and perpetual motion in Pierre's thought. His ideas on the subject were elaborated in the second part of the "letter on the magnet," where he described another method of achieving perpetual motion "by means of the virtue or power of this stone." The device involved in this shows certain similarities with a mechanism described in 1271 by Robert the Englishman, who was mentioned above. Robert seems to have shared Pierre's interest in making a wheel to revolve in time with the heavens. But Robert's machine did not involve a magnet, and historians have suggested that he was describing an early but unsuccessful attempt to make a weight-driven clock. It certainly included a wheel that would rotate once every twenty-four hours: "Clockmakers are trying to make a wheel which will make one complete revolution for every one of the equinoctial circle, but they cannot quite perfect their work. But if they could, it would be a really accurate clock, and worth more than an astrolabe or other astronomical instrument for reckoning the hours. . . . The method of making such a clock would be that a man make a disc of uniform weight in every part so far as could possibly be done. Then a lead weight should be hung from the axis of that wheel so that it would complete one revolution from sunrise to sunrise."[10]

Pierre de Maricourt was the first European author to discuss magnetism in any detail, but the first reference to the magnetic compass in Europe is to be found in a work written by Alexander Neckam about seventy years earlier (c. 1200). But although the compass had been in use for some time before Pierre wrote, it seems likely that he made significant improvements to the two versions of the instrument he discussed. His wide practical interests were known to Roger Bacon, as Bacon mentioned in a work called *De secretis,* a few years before the "letter on the magnet" was written; and further references occur in three *Opera* by Roger Bacon that were written around 1266–7.

In the first of these works, after describing the armillary sphere, Bacon commented that a "faithful and magnificent experimenter" (presumably Pierre) was trying hard to make such a sphere "out of such material and by such a device that it will revolve naturally with the daily rotation of the heavens."[11] He supported the idea that this was possible by saying that after all, the tides move perpetually in time with the motions of the moon and sun. These comments again call to mind the wording Robert the Englishman used to discuss wheels that would rotate once every day, keeping in step with the stars.

Clocks and Cathedrals: 1140–1280

It is impossible to say whether the views of Roger Bacon and his colleagues were shared by the men who made the first weight-driven clocks around 1300. But it is clear that Roger, Pierre, and Robert all thought about clocks in terms of a celestial prototype: the motion of the sun and stars was the only perfect clock, and any man-made timepiece ought to imitate the heavens not only in precision of movement but also in an untiring and continuous motion.

Men with practical experience would without doubt recognize that this was an unattainable ideal. But even ideals that are known to be unattainable are capable of influencing action.

This one certainly was, and its influence perhaps accounts for the restless inventiveness of the time. There was a desire to test nature and to see how near to the celestial prototype human devices could come.

The inventiveness of the people who devised the clocks and textile machines of the period 1250–1350 recalls the prolific inventiveness of the cathedral builders in the previous hundred years, and recalls too their attitude of never being content even with the stupendous achievement of a Chartres or an Amiens. If the clockmakers inherited their outlook, it does not seem surprising. The thirteenth-century craftsmen shared to some degree in the Roman view that "architecture is divided into three parts: building, clocks and machines."[12] This is shown by the sketchbook of Villard de Honnecourt, which was compiled over a period of perhaps thirty years from c. 1235, and which Pierre de Maricourt could, just possibly, have seen. Apart from the many drawings related to architecture and sculpture, Villard sketched a perpetual-motion machine, some gadgetry probably connected with the mechanism of a water clock, and the large masonry case in which such a clock would be housed. The latter is labeled unmistakably: "this is the housing for a clock."[13]

With this community of interest between clockmakers and cathedral builders, one may well inquire in the same terms about the reasons for the inventiveness of both groups. If some clock-makers saw the sun and stars as a kind of celestial prototype for their machines, did the cathedral builders have the stimulus of any similar imaginative ideal? In the previous chapter it was said that there was some stimulus from mutual competition between one building project and another. But were there also more noble, perhaps religious, ideas that would work within the imagination of the builders rather as the idea of a celestial clock played on the thoughts of Pierre de Maricourt?

Reasonably clear answers can be given to these questions because statements about the significance of church buildings can be found in the liturgy for the consecration of a new church, in the writings of Abbot Suger in the 1140s, and also in the sculpture that adorned the buildings themselves. And this evidence makes it plain that every church was understood partly as a symbol of the New Jerusalem that St. John the Divine had seen in a vision descending from heaven. This vision was described in the Bible, in the book of Revelation whose great significance for many people of this period has already been noted. St. John's vision portrayed a new order of things that would come about after the end of the world as we now know it, and after the Last Judgment. Church buildings themselves were designed in part to represent that vision, and many French cathedrals of the twelfth and thirteenth centuries had a sculpture depicting the Last Judgment in pride of place above the main western entrance. A visitor to the cathedral would thus be presented with an image of this highly disturbing event before he or she symbolically entered the heavenly Jerusalem as represented by the building itself.

The liturgy for the consecration of a new church carried many other layers of meaning also. The church could be related to the historic Jerusalem and the temple Solomon had built there.[14] But passages were read from Revelation that explicitly described the New Jerusalem, and the new building was marked with twelve crosses, corresponding to the twelve foundation stones of the heavenly city.[15]

The writer of Revelation had described the New Jerusalem in some detail, and his account suggested several images church architects could use, such as the great height of walls made "of pure gold, transparent as glass" (Revelation, xxi, 12, 18) and glittering "like some precious jewel or cut diamond." Such an effect could be created in a cathedral by the glowing reds and golds of translucent stained glass.

A significant comparison is possible at this point. The cathedral as a stone embodiment of the New Jerusalem was a scaled-down, material model of heaven, just as clocks were a scaled-down representation of the visible cosmos. Abbot Suger in 1140 said that his church was built "in likeness of things Divine," and Pierre de Maricourt could have said the same in 1269 about the magnetic clock he proposed. Thus there was a sense in which the cathedral builders, like the clockmakers, had a celestial prototype.

There was here an echo of the teachings of Plato, who had said that all material objects were merely corrupt and distorted representations of "eternal ideas" found only in heaven. Any triangle drawn by a man could only aspire to be a rough sketch of the idea of a triangle; any motion on the earth could only approximate to the uniform circular motion the heavenly bodies exhibit; and it would follow that the earthly Jerusalem that the crusaders had tried to save—and the great churches of France—were all of them imperfect models of the heavenly City of Jerusalem that the Bible had described.

Light, Geometry, and Architecture: 1140–1280

Plato's thought was very influential in the twelfth and thirteenth centuries, particularly in the highly imaginative interpretations given to it by such later writers as Plotinus (c. A.D. 250), Dionysius (c. A.D. 500), and St. Augustine (A.D. 354–430). The "neo-Platonic" philosophies developed by these men influenced development of the technical arts in at least two ways: first, through the stress Plato had laid on mathematics, and second, because the relationship between terrestrial and heavenly bodies, which neo-Platonists envisaged, could be used to support a theory of astrology and magic.

The latter point can be explained by recalling that the stars belonged literally to "the heavens" and had quasi-spiritual qualities,[16] but terrestrial matter was intrinsically corrupt, dead, and without "form." It was indeed only through power that came from a higher order that life and motion was possible on earth at all. The tides in the sea were driven by the action of the moon; compass needles were drawn to the pole star; and all life depended on the light of the sun. Indeed, light was the most powerful of all the forces that came from the heavenly bodies.

The most important exponent of this point of view was Robert Grosseteste, a teacher at Oxford early in the thirteenth century and Bishop of Lincoln from 1235 to 1253. Grosseteste studied mirrors, lenses, and light in a thorough and almost scientific manner. He was not alone in this, and not long after his death in 1253, spectacles with convex lenses came into use in Italy. But Grosseteste's study of light was inspired by a near-astrological view of its powers. For he believed in a "light-power or virtue" that was directed on to the earth from the heavens. This special radiation "was the basis of all bodily magnitude and of all natural operations, of which the manifestations of visible light was only one. . . . The concentration of light-power or virtue from the heavens on to the earth . . . caused the differences in climate observed in different parts of the globe, the growth of plants and animals, the transformation of one of the four elements into another, the astrological influences varying with the configurations of the planets, and the rising and falling of the tides."[17]

Pierre de Maricourt was also interested in light, and in particular its reflection by concave mirrors. What is more, he spoke of a "virtue" directed on to the earth from the sky. This "virtue" was not light—not in the ordinary sense—but the power that caused a suspended lodestone or compass needle to point always to the north. So Pierre remarked that the "poles of the lodestone derive their virtue from the poles of the heavens. As regards the other parts of the stone, the right conclusion is, that they obtain their virtue from the other parts of the heavens"[18] But the magnetic virtue was discussed in very much the same terms as those which Robert Grosseteste used to describe light, and Pierre, who must have known of Grosseteste's work,[19] evidently thought about light and virtue as similar kinds of phenomenon. So Pierre said that one experiment on "the virtue of lodestone, I will explain in my book on the action of mirrors"[20]—presumably because the two things were for him related subjects.

Other metals and minerals besides lodestone were supposed to receive different virtues from various heavenly bodies, and there was a close link between the idea of celestial forces influencing magnets and the tides and the ideas of alchemy and astrology. The sun was associated with gold, the moon with silver, and the five known planets with five other metals. It was through these and many other connections as well as through light and the magnetic virtue that all kinds of remarkable things, including perpetual motion, might be effected by man.

However, what was of greater importance for the development of the practical arts in the twelfth and thirteenth centuries was the interest then shown in mathematics. This was encouraged by what people had read of Plato's *Timaeus,* especially at the school of Chartres, where Platonism was a direct stimulus to mathematical learning. For Plato had emphasized that the universe was mathematical in its structure (*Timaeus,* 31, 32, 36), and it followed that mathematics was needed if one was to understand nature. This attitude was interpreted by St. Augustine in such a way that the simple numerical relationships in which Plato had been interested took on a theological significance. Three was the number of the Trinity and so of God; four stood for matter; seven, the sum of three and four, signified mankind; and twelve, the product of three and four, was the number of the apostles and so of the Church. Number was impressed upon everything and was a sign of the divine wisdom.

Mathematics, of course, must always play an important part in the art of building, because it is necessary for surveying and for setting out the plan of a building on the site. But Abbot Suger's abbey church of St. Denis and the contemporary cathedral at Sens were pioneers of the intensely geometrical Gothic style of architecture. The artistic effect of the Gothic style was always to stress line rather than bulk, and such architects' drawings as survive from the Gothic period always stress the linear characteristics of designs. One writer of 1140–50 said that the work of a building craftsman consisted of "practical geometry . . . the forming of lines, surfaces, squares, circles and so on in natural bodies."[21]

Now it is interesting to note that the new Gothic architecture was being developed in France at just about the time when the first Latin translation of Euclid's great work on geometry arrived there. These translations had been made from Arabic versions of Euclid, and before this, medieval Europe had been largely ignorant of the mathematical achievements of the ancient Greeks.

At the cathedral school at Chartres the rediscovered theorems of Euclid were carefully studied, and it was there that they were first taught. Chartres, in fact, became for a time the most important school of mathematics in the West. This was at least partly because it was already a place where Platonic thought received close attention, including the first part of the *Timaeus* and

Augustine's ideas about numbers. For Plato's emphasis on mathematics greatly stimulated interest in geometry.

Master masons and architects who were working in the region would naturally share this enthusiasm, not only because of the practical value of geometry in their work but also because some of them were probably taught by masters from Chartres. The cathedral schools, which were the predecessors of the universities soon to be founded at Paris and elsewhere, often enrolled pupils who intended to become skilled craftsmen. Such were the architects of the cathedrals—literate men who knew Latin and who were quite capable of keeping in touch with academic geometry; they were men who could write about mensuration and surveying: one anonymous French author of the thirteenth century described methods used by masons for finding the areas of triangles, octagons, and circles.[22]

The sketchbook of Villard de Honnecourt contained similar instruction in geometry; he wrote about what "the art of geometry commands and teaches" with a philosophic conviction suggestive of Platonism. But it was on a severely practical level that Villard was most concerned to discuss geometry, for the craftsman constantly needed to use it in order to ensure that the objects he made were precise in shape and dimensions.

The method can be illustrated by reference to the screw jack Villard drew in his sketchbook. It is so rough a drawing that one wonders whether he could have made such a machine with any reliability. However, instructions about cutting screws that were added to the book a little later show that this sort of job could be done carefully and accurately.

In the example given, the screw was to be made from a cylindrical wooden rod and its pitch was to be one-third of the diameter of the rod; a sketch showed how intervals of this length were to be marked off (figure 10). The technique was to mark the diameter of the rod as a length of another piece of wood, then use the standard geometrical construction to divide a line into three equal parts. With these three equal lengths clearly marked on the piece of wood it became a purpose-made gauge. Then to complete the job, a straight line was drawn along the length of the rod, and the gauge was used to mark off the required intervals. Finally, a length of cord was wound in a spiral that intersected these points. So by drawing along the line of the cord, one obtained the line of the screw thread directly.

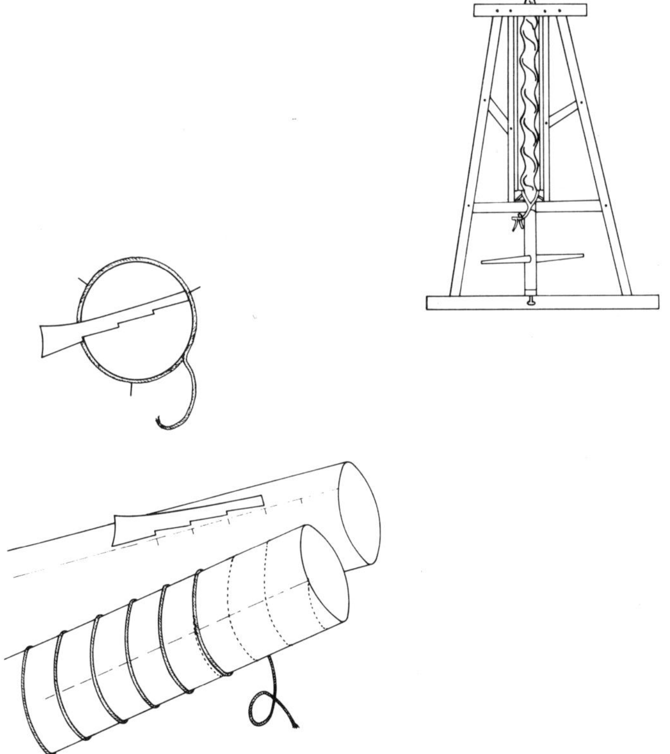

Figure 10
Diagram after Villard de Honnecourt's sketch of a screwjack. (Drawn for the author by D. M. Farrar.)

Another practical application of geometry was in the drawing of sea charts. The merchant seamen whose ships plied from Venice and Genoa set course by means of the magnetic compass and roughly estimated the distance they traveled on each bearing. Sailing directions for all ports, islands and headlands were compiled, specifying the direction to steer from, say, Genoa, and how far to sail on that course. In the thirteenth century, not long after Euclid's geometry had come to be taught in Italy, it was seen that these sailing directions could be represented geometrically. With two circles, each marked with sixteen points of the compass, sailing directions for the whole Mediterranean could be illustrated on a single chart (see figure 11). Such charts were the first maps to be made on the basis of direct (if approximate) measurements, and the earliest that survives dates from around 1275.

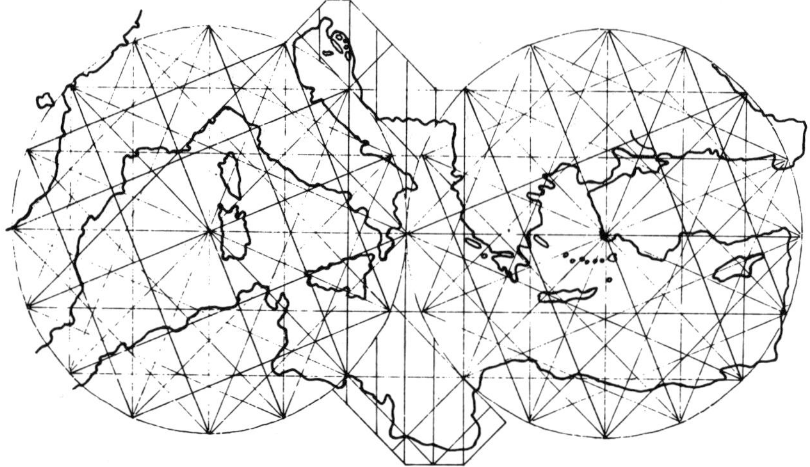

Figure 11
Geometrical method for drawing a sea chart, approximately as used on the *Carta Pisana* of c. 1275, the earliest map based on direct measurements.

In the design of buildings, geometrical procedures comparable with the two preceding examples were used. The best-known technique was the system of construction lines based on the geometry of the square, by the which the height of a cathedral vault and other dimensions were worked out. This method of "*ad quadratum* design" was probably used in building most of the major thirteenth-century churches (see figure 12). Evidence for its application at Amiens, Beauvais, Cologne, and Westminster is fairly clear.[23]

A comparable approach to a different problem is provided by the square and its diagonals that Villard de Honnecourt suggested as an aid for drawing a man's face in correct proportion. This too is an *ad quadratum* method. Other aspects of the geometry used by builders at this time have come to light from studies of drawing instruments, and the equipment employed in setting out the lines of a building on the ground or marking stones for cutting by the masons. These instruments differ somewhat from those familiar to modern draughtsmen (figure 13) and were not used in the same way. Architects made relatively few drawings on paper or parchment, and these were not to scale. Sometimes, though, they sketched out ideas on the walls

Figure 12
Ad quadratum design. Construction lines for drawing a man's face, as suggested by Villard de Honnecourt, c. 1235.

of the building they were erecting, as at Ashwell Church, in Hertfordshire (England). To a large extent, it would seem, the designer of a building worked out details of the plan in the process of surveying and setting it out on site, no doubt with much repetition and correction. Details of piers, arches, and windows were regularly drawn out full size on large, carefully smoothed "tracing floors." Documents suggest that these existed at most major building sites, and a good deal is known of one at Strasbourg. Two have survived in England, at Wells and York. In each case, the floor surface is of plaster of paris, and is covered with lightly inscribed lines, which are the superimposed remains of many drawings. The visible lines on the Wells floor show only a few recognizable details of shapes of stonework and a doorway, but they include numerous curved lines drawn with compasses.[24] No modern draughtsman would need his compasses so frequently, and this is clear evidence that Euclidian geometry rather than measurement was the basis of drawing.

The smaller drawing instruments of the time included set-squares as well as compasses (figure 13). These were needed for drawing out patterns and templates that masons could employ to check the shapes of stones they were cutting, as well as being used on the tracing floor. For setting out the plan of a structure on site, a long measuring rod or "Great Measure" provided a base length from which other dimensions could be constructed

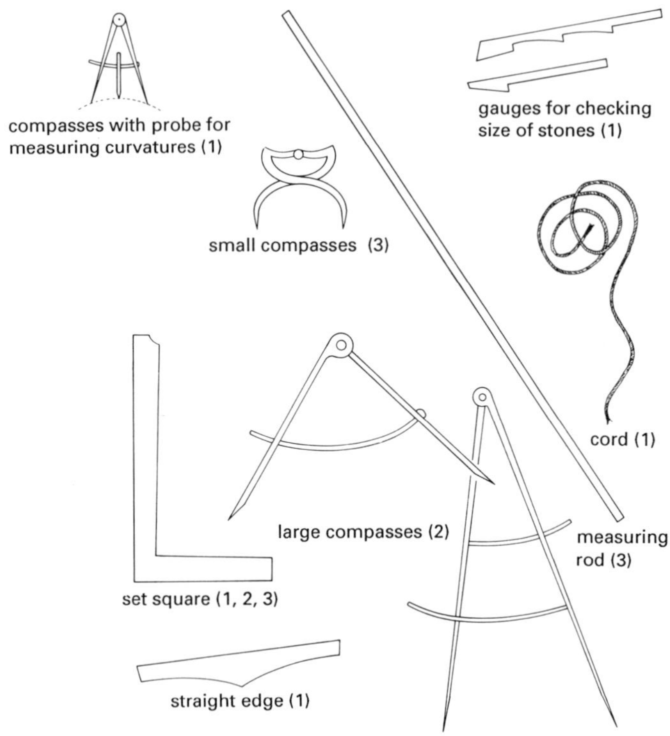

Figure 13
Masons' drawing instruments.

(Sources from which these are known: 1. notebook of Villard de Honnecourt; 2. life of St. Offar, thirteenth century; 3. tombstone of Hugh Libergier, mason, 1263. Drawn for the author by D. M. Farrar.)

geometrically. A length of cord attached tautly to a fixed point was used on the building site to mark out arcs of large radius.

The set-square used by architects and stone masons was L-shaped, with the arms tapered in such a way that the outer and inner edges of the square could represent two right-angled triangles with slightly different proportions (figure 14). These triangles then formed the basis for a series of geometrical constructions that were used in the design of church buildings. B. G. Morgan, who has recently discussed the use of this square in great detail,[25] believes that its earliest application in England was at Bristol Cathedral in 1298; it had probably been invented in France in the twelfth century and was illustrated by Villard de Honnecourt in a part of his book compiled about 1250. Its introduction at Bristol may be connected with a man signifi-

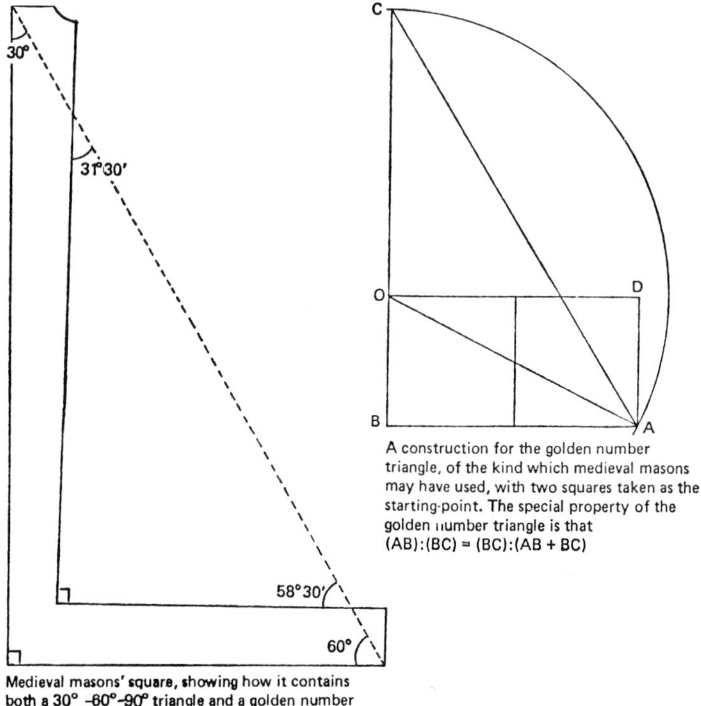

Medieval masons' square, showing how it contains both a 30° –60°–90° triangle and a golden number triangle

A construction for the golden number triangle, of the kind which medieval masons may have used, with two squares taken as the starting-point. The special property of the golden number triangle is that (AB):(BC) = (BC):(AB + BC)

Figure 14
A set-square of the type used by masons in the thirteenth century for marking out pieces of stone before they were cut to shape and for setting out the plans of buildings. The square could be used to draw either of the two kinds of triangle indicated, and the ratio (AB) : (BC) was frequently reflected in the proportions of the finished building.

(Drawn for the author by D. M. Farrar based on illustrations in B. G. Morgan, *Canonic Design in English Medieval Architecture,* Liverpool, 1961, by permission of Liverpool University Press.)

cantly called William the Geometer who was buried there soon after 1300.

For the cathedral builders, then, mathematics was a basic discipline; but of course other aspects of thirteenth-century thought could influence them as well. In particular, this was a period when the works of Aristotle were receiving close attention, having not long before been reintroduced into Europe from Arab sources. This interest in Aristotle coincided with and reinforced a growing emphasis on the orderly classification of knowledge, and in the middle years of the thirteenth century such authors as Albertus Magnus and Vincent of Beauvais embarked on ambitious encyclopedic works that discussed the whole of contemporary knowledge according to carefully thought-out systems of classification. The culmination of these studies was the work of St. Thomas Aquinas, a pupil of Albertus Magnus. Aquinas, who died in 1274, provided a masterly synthesis of Christian doctrine and Aristotelian thought. St. Thomas, who saw his vocation as the ordering of human thought, once remarked that an "architect" was a man who knew how things should be ordered and arranged, and that the word could be more appropriately applied to a philosopher than to a builder.[26]

In practice master masons rarely called themselves architects at this time, but in some respects their work did reflect the ordering of knowledge that the Aristotelian philosophers had effected.[27] For example, the numerous sculptured figures in thirteenth-century cathedrals were often arranged to illustrate the classification of knowledge put forward by encyclopedists such as Vincent of Beauvais. Vincent described the practical arts in terms of a number of limited categories such as alchemy, metalworking, and building. In the north porch of Chartres Cathedral there are carved figures representing these same three practical arts—metalworking is represented by a man with an anvil; building by a man holding a set-square and compasses; and alchemy by the figure of a *magus* who holds a scroll marked with the signs of the planets and metals and who has a dragon at his feet. Other figures represent medicine and painting, and a man with a plow stands for agriculture.

While the world of learning came to be dominated by Aristotelian thought, with its fondness for classification and the rational ordering of knowledge, artisans and craftsmen continued to operate with ideas that owed more to neo-Platonism than to Aristotle. They continued to discuss astrological influences, light, and magnetic virtue, none of which could fit easily into an Aristotelian frame of thought, and a linked idea was the notion that both clocks and cathedrals were images of heavenly prototypes. The effort made by craftsmen to match these prototypes is a particularly clear example of an idealistic tendency in technology, and can help us to understand the surge of inventive activity that occurred at this time.

Invention is always an act of the imagination. One reason why Europeans proved to be so good at it in the thirteenth century and since is that European culture offered much to stimulate the imagination, notably in the symbolism associated with cathedrals and clocks. There was also the flux of ideas derived from Greek thought and the technical knowledge of Islam and the East.

All this helps us to explain the "high technology" of the time, if that phrase may be used to describe the cathedrals and clocks. But there was also the more practical aspect of innovation, which found its biggest achievements in the new textile machines of the thirteenth and fourteenth centuries. These provide some corrective to any overemphasis on the symbolism of the cathedrals. Medieval artisans were obviously capable of working within a severely practical frame of reference for much of the time, and we should not suppose that they were always entranced by a highly colored symbolism. But the machines developed for use in this more practical sphere of activity were far simpler and less technically ambitious than anything involved in clockmaking. The silk-twisting machine was, indeed, the only really complex industrial mechanism.

So this was a period when much of the most advanced technology was stimulated by imaginative and not material motives. The historian does well to remember that "to discover economic causes is to some degree a craze with us, and sometimes leads us to forget a much simpler psychological explanation of the facts."[28] With thirteenth-century technology, the psychological explanation is a matter of symbolism and of ideas about celestial objects that seem remote and strange to twentieth-century thinkers. But these ideals provoked an enormous thrust forward in the practical arts and helped to create the state of mind in which people could envisage technical improvement and work to achieve it.

3

Mathematics and the Arts: 1450–1600

The Age of the Printing Press

The first books to be printed in Europe were probably produced about 1448 at Mainz on the Rhine. The printing office responsible had been set up by a certain Johann Gutenberg, a goldsmith, and he is conventionally regarded as "the inventor" of printing by moveable type.

Printing had been in widespread use for a considerable time before Gutenberg, first in China and then in the West. But only pictures and playing cards had previously been produced in Europe, and these were printed from blocks, without the use of type. What was invented some time before 1448 was metal type, pieces of which could be assembled to represent the page of a book. When sufficient copies had been printed, the type could be used again, rearranged to represent quite different words and sentences. The system depended on having large numbers of pieces of type, strictly standardized in size and shape. This requirement was met by making the type from a tin-lead alloy similar to pewter and casting it in a copper mold such as pewterers used. Probably several people had experimented with this technique before Gutenberg produced the first printed book.

The idea that Gutenberg was the sole inventor of printing grew up at the end of the fifteenth century, at a time when people had come to think of the work of any great artist, poet, or inventor as the product of special creative genius that the majority of ordinary men did not possess. The idea that men could be truly creative was a new and exciting idea at this time. Creation had previously been thought of as the prerogative of God; now it was seen to be an activity in which mankind could share. But it was not just anybody who could partake of this godlike attribute—only rare individuals such as Michelangelo.

When ideas such as this were applied to thinking about technical change, they immediately led to the idea of invention as a unique act of creation, rather than as an evolutionary process to which many minds might contribute. Stress on the individual as creator was very characteristic of Renaissance art, and it affected technology also, because at this time, artist, architect, and engineer were often one and the same. An example is Filippo Brunelleschi (1377–1446), one of the pioneers of Renaissance architecture. He invented hoists for lifting stone and boats for carrying it, as well as special techniques used in constructing the dome of Florence Cathedral. His inventions were rewarded by prizes, and in one case he was granted a patent of monopoly for a three-year period.[1] This was in 1421, and is the earliest known instance of a patented invention. The practice of issuing such patents eventually gained wide acceptance—it was formally adopted in Venice in 1474 and in England in 1623—and this gave even more emphasis to the role of the individual inventor.

As individual invention had been recognized in this way during Brunelleschi's time, it should come as no surprise that when printing came into general use a little later, people began looking back to its origin in the hope of identifying an individual of genius who invented it. Gutenberg's name became well known; and his early success with printing by moveable type seemed to be an outstanding instance of an invention that could be credited to a named individual.

The idea of invention soon came into such prominence that it was taken to represent the essence of technical progress. Writers made lists of inventions, and very often special attention was given to firearms, printing, and the magnetic compass as the greatest of recent inventions. A certain Polydore Vergil produced a book of inventions in 1499; another was written by Jean Taisnier in 1562. Jean Fernel in 1548 and Francis Bacon after 1600 also used great inventions as indicators of technical advance.

Today, most historians of technology would reject this attitude, although it has influenced many popular works on the subject. As one historian has remarked, it is misleading to present invention "as the achievement of individual genius, and not as a social process."[2]

The social process which led to the "invention" of printing consisted partly in the development of earlier forms of printing;

partly in the numerous experiments with type that several people seem to have made in the 1440s; and partly in a widening literacy and a growing demand for books. All these developments were most marked in Italy, although they could also be observed in the Low Countries, in Paris, and in a few German cities, like Cologne, Augsburg, and Mainz. That the ground was already well prepared before Gutenberg's time is shown by the rapidity with which printing by type came into use. Within thirty years of the establishment of Gutenberg's press at Mainz, there were 236 printing presses in Italy, 78 in Germany, and 68 elsewhere; by 1500, 532 presses had been established in Italy, 214 in Germany, and 304 elsewhere. Venice was the most important center; in 1500 there were nearly as many presses in Venice as in the whole of Germany.[3]

The demand for books was one result of the complex cultural change for which the term "Renaissance" has already been used. Beginning in Italy around 1400, the Renaissance brought a great revival of scholarship and learning. The printing press grew up in the service of Renaissance scholarship, and the art of printing was a Renaissance technology. Thus much of its early development took place in Italy. The demand for printed books and the early development of block printing began in Italy, and it was there that printing by type proliferated most rapidly. This is a point to be emphasized, despite the undoubted success of Gutenberg of Mainz as the pioneer of book production.

Movements in Technology

Some historians of technology see the Renaissance as the period in which there first emerged the restless, technical inventiveness that so strongly characterizes Western culture.[4] They point out that it was at this time that the pursuit of novelty in technology was first recognized and honored, notably in the patents granted to Brunelleschi in 1421 and in the books that listed path-breaking inventions such as firearms and printing.

It is certainly true that these developments mark a new *awareness* of invention. It is also true that the appearance of so much material in printed books helped to disseminate an interest in technical novelty very widely, and probably stimulated other inventors.

But for three of four centuries before this, as previous chapters have shown, Europeans were prolific inventors and resourceful adaptors of techniques borrowed from other cultures. What happened during the Renaissance was that this inventive spirit achieved self-consciousness. The pursuit of novelty and technical improvement became a recognized goal and, indeed, an established ideal.

A further point to make about the history of invention in Europe is that it can usefully be seen as a succession of *movements* each characterized by the exploration of its own cluster of ideas and techniques.[5] The invention of printing by movable type and its very rapid expansion from a few pioneer presses in 1450 to over a thousand by 1500 illustrates very clearly how a movement in technology can be established by a series of inventions and then grow rapidly, both through further technical development and by wider commercial exploitation.

Previous chapters have stressed three other clusters of innovations that can be seen developing as distinct movements. These were the refinements of the flying buttress and cathedral vaulting techniques between the 1090s and the 1270s; the application of water power to new uses such as crushing ore; and the development of the weight-driven clock.

In a more comprehensive history of the period, several other movements in the practical arts would need to be mentioned, some relating to improvements in agriculture—the most fundamental of all technologies—and some concerned with the spread of new spinning, weaving, and silk-processing techniques. There was also a remarkable movement in the technology of warfare, which saw the rapid improvement of guns from their first introduction about 1300. By 1500, not only had the cannon become a formidable weapon for use in sieges and on ships, but also the matchlock musket was taking recognizable shape after a long evolution from more primitive forms of handheld weapons.[6]

Each of these movements in technology was characterized not only by a group of innovations but also by a characteristic institutional and social context providing for the development of skills, the spread of information, and the reinforcement of shared goals and values. Often ideas spread informally, through the travels and mutual contacts of individual craftsmen such as Villard de Honnecourt and Pierre de Maricourt (chapter 2). However, a clear example of the involvement of formal institutions is the development of water power within Cistercian monasteries (chapter 1).

In the remainder of this chapter we are concerned with two movements in the practical arts centered on the independent city-states of northern Italy, both concerned with applications of mathematics. The achievements of the first, whose focus was in Florence, were related to ideas about architectural proportion and techniques for making drawings to scale. The other, developing rather later in the environs of Venice, was more concerned with machines. Both benefited from Renaissance interest in translating, studying and interpreting what books survived from the ancient Greek and Roman civilizations. For example, enthusiasm for mathematics was encouraged by a fragmentary knowledge of Pythagoras and by detailed study of Plato. Interest in machines was encouraged by what was discovered about later Roman and Alexandrian authors, especially Vitruvius, Archimedes, and Hero of Alexandria (figure 15).

Among these various books, one that greatly attracted Renaissance scholars was compiled about 30 BC by the architect Vitruvius. This not only discussed architecture but described machines and the various engineering techniques that architects needed to use. During the fifteenth century, architects and scholars studied the remains of Roman buildings, many of which survived in Italy, and interpreted the writings of Vitruvius in the light of what they saw. Some aspects of engineering came into these scholarly, almost academic studies, and it is striking how often artists and scholars became involved in engineering, often in connection with building works, fortification, or water supplies.

Some historians have spoken of these men as "artist-engineers" or "artist-technicians,"[7] and to a large extent, the movements in technology with which we are concerned in this chapter are associated with such people.

Mathematics and Drawing in Florence

In Florence, from the 1460s, a group of scholars informally constituted themselves into an academy modeled on Plato's academy in ancient Athens. Their outlook was literary rather than mathematical, but they were well aware of Plato's emphasis on number and proportion as a means of understanding nature.

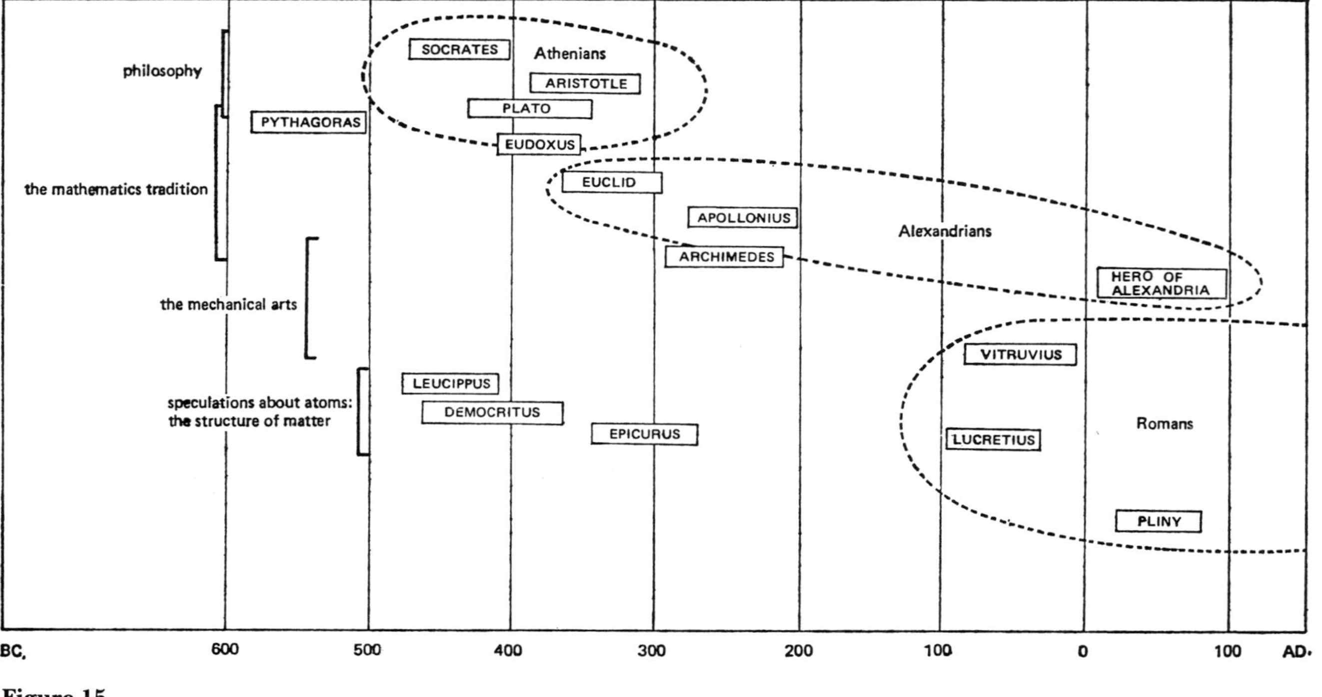

Figure 15
Time chart of Greek authors mentioned in chapters 3 and 4.

It is significant that Leone Battista Alberti (1404–72), the leading Florentine architect of the time, was one of the members. Alberti had studied the ruins of ancient buildings in Rome and pioneered the design of new buildings in a Roman style. This new architecture was not entirely compatible with the builders' mathematics used by medieval designers, and Alberti belonged to a generation that was exploring new approaches. The emphasis on number and proportion he found in Plato and neo-Platonic writings was an important stimulus for this.

But when Alberti came to write a book on architecture—the first on the subject to be printed, though only after his death—he showed himself to be an artist-engineer as well as an exponent of mathematical proportion. He included several chapters on engineering in relation to building, civic design, and public works. Some of the most interesting were on water supply[8] and on the provision of water to drive mills.

In his chapters on architecture proper—the design of houses, churches, and other buildings—Alberti made it clear that one aspect of builders' mathematics that particularly concerned him was the question of proportion. The basis of his ideas was an analogy between the beauty of a fine building and harmony in music. Musical harmony was believed to be governed by mathematics, and "the same Numbers, by which the agreement of Sounds affects our Ears with Delight, are the very same which please our Eyes and our Mind."[9]

This is where the growing Florentine enthusiasm for neo-Platonism was relevant, because ideas about music and proportion were connected with the mathematical speculations of Pythagoras and Plato. Some of these philosophers in the ancient world had made experiments with the strings of musical instruments. By comparing the sounds emitted by different lengths of a string under constant tension, they discovered that musical harmonies corresponded to simple ratios in the length of the string. For example, if the length of the string were halved by a bridge, it produced a note exactly one octave higher than it had done to start with. If only two-thirds of the string could sound, the note would be higher by an interval of a fifth.

In this way, the Pythagoreans found that the basic notes in their musical scale were all related by simple ratios of whole numbers. They regarded this discovery not just as applying to music but as giving a clue to the harmony of the whole universe.

For this last reason, and because Alberti thought that the arts should reflect the cosmic harmony, nature being the "greatest Artist," he believed that these numbers were applicable in architecture. This might mean, for example, that the rooms in a house should have sides whose measurements were in the ratio of three to two, with the height equaling the width. Or alternatively, a room might have the measurements of a double cube: that is, it would be two units long, one unit high, and one wide.

In his book, Alberti had commented that the human body offerred an example of good proportion. Others pursued the point and some took measurements in order to investigate the proportions of human bodies in detail. One of those who did this was Albrecht Dürer, a German from Nuremberg who had visited Italy several times and had come to share some of the enthusiasms of the Italian artists. It is worth noting that artists were led by these interests to make careful *scale drawings* (figure 16) and in some instances to use a series of scale drawings to construct geometrically accurate *perspective* views. Scale drawing and perspective were important innovations of the period, used by architects as well as painters.

Although it was a long time before engineers used these techniques systematically, scale drawing was ultimately of great importance in enabling machines and structures to be more thoroughly thought-out (in terms of design) before construction began. This gave engineers a larger scope to explore new ideas.

Another context in which measurement and scale drawing had important applications was in geography and mapmaking. This can be seen in the work of Paolo Toscanelli (1397–1482). A contemporary and colleague of Alberti, he was outstanding among the learned men of Florence for his work in astronomy. The contemporary interest in using measurement as a tool for investigation is shown by the accuracy with which Toscanelli recorded the altitude of the sun at different seasons of the year, and by his efforts to decide how large the continents of the earth should be when represented on a map. He was rightly convinced that Asia was much larger than people had previously imagined, and he questioned travelers about journey times and distance in order to try and form an estimate of its size.

His ideas on this subject reached the ears of a young Italian seaman, Christopher Columbus (1451–1506), who had sailed to Ireland and perhaps even as far as Iceland. In those countries, there was rumor of land even further to the west—in fact,

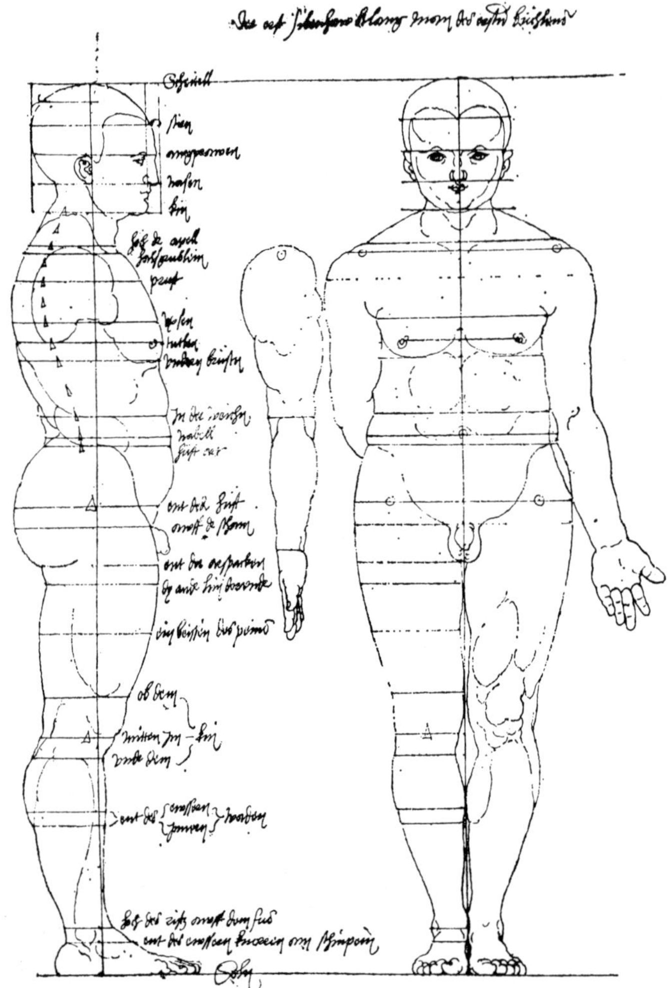

Figure 16
Albrecht Dürer's drawing "Proportion figure in profile and full face," c.1523.

(Courtesy of the Fogg Art Museum, Harvard University, Cambridge, Massachusetts, bequest of Charles A. Loeser.)

Greenland. In 1478, Columbus wrote to Toscanelli about the prospects of reaching Asia by sailing westward into the Atlantic. The reply was encouraging, and with his letter Toscanelli sent a chart showing the islands of Japan and the East Indies in relation to the coasts of Europe. From surviving fragments of the correspondence, it is known that Toscanelli thought the distance between Europe and the East Indies to be about 130° of longitude. From this, and from a figure for the earth's circumference that turned out to be too small, Columbus deduced that Japan should be about 3,000 miles (5,000 kilometers) to the west of Spain—that is, about the distance that America proved to be. When he finally reached America in the 1490s, Columbus persisted in thinking that the coasts and islands he was exploring were part of Asia.

Toscanelli, like Alberti, had connections with the Platonic academy in Florence. But while the academy's revived neo-Platonism influenced studies of the natural world, which with their measurements, maps, and scale drawings were almost of a "scientific" character, there were other tendencies in neo-Platonism as well.[10] Plato's view of the world had been poetic and even religious. Numbers and geometrical shapes could be given a mystical significance. Discovery and invention could seem to be the product of inspiration, not of deduction and experiment. Moreover, astrology was given intellectual justification by the neo-Platonic worldview, because the stars and planets radiated light and "virtue," which was a vital influence on terrestrial life. As we saw in chapter 2, ideas of this sort were already circulating in the thirteenth century, but now they were reinterpreted and given fresh impetus. Toscanelli, the leading astronomer in Florence, was also the leading astrologer, and Alberti was not beyond consulting astrologers as to the best time for laying the foundation stone of a building.

It needs to be clearly understood, therefore, that although the intellectual movement described in this chapter saw the birth of rational techniques of measurement and drawing, and encompassed men with a clear grasp of engineering problems, this was not the materialistic, disenchanted rationalism of the modern worldview. Nature was still felt to abound in hidden forces and quasi-spiritual influences, and what we now call technology still had something of the quality of magic.

Agriculture and Architecture in the Venetian Republic

A century after Alberti, an architect with similar ideals about the application of mathematics to building was Andrea Palladio (1518–80). Much of his career in architecture was spent in building houses for gentlemen from Venice, whose wealth had its origin in merchant shipping.

Over the years, Venice had grown prosperous by trade with countries in the eastern Mediterranean that provided a vital link between Europe and places farther east. But in Palladio's time, the commercial preeminence of Venice was threatened. Other countries now had faster ships that were cheaper to operate. By the 1530s, too, the new sea routes opened up by voyages of exploration around the coast of Africa were beginning to divert trade away from the Mediterranean. Portuguese, Spanish, and Dutch ships sailed the Indian and Atlantic Oceans, carrying cargoes that had once been the Venetians' staple trade. The Venetians realized that to maintain their prosperity they would have to rely less on trade and pay more attention to the development of their own internal resources. Thus the sixteenth century saw the growth of cloth manufacture and glassmaking in and around Venice, and there were already many thriving printing shops. But most significant of all, the Venetians embarked on a vigorous policy of agricultural development in their hinterland. Between the 1530s and about 1610, there was a series of land drainage schemes and projects for the control of flooding from rivers. Waste land was enclosed, improved strains of wheat were introduced from Turkey, and cereals were grown in vastly increased quantities.

In 1556, the Venetian senate created a Board of Uncultivated Properties that administered state subsidies for land reclamation and coordinated the efforts of individual landowners. Marshes were drained and brought into cultivation, while canals and embankments were constructed to prevent the overflow of rivers. Merchants from Venice, tired of the risks of maritime trade, sought more secure investments by buying estates and becoming gentlemen-farmers.

This agricultural revolution, as it can appropriately be called, was of much more than local significance, and to understand its context we need briefly to consider the state of agriculture throughout Europe. Beginning just before 1300, static or falling agricultural production became a problem in many parts of

Europe. This was one of the reasons why the brilliant achievements of the thirteenth century did not lead to a phase of continuing technical development. During the fifteenth century, the revival of learning and of the practical arts was sustained by an economic revival, of which developments in agriculture were an important part. By the sixteenth century, agricultural improvement was in progress in many parts of Europe. For example, in southern Spain, where some crops depended on irrigation, a series of large dams was built to allow more land to be cultivated. One of these, near Alicante, was completed in 1594, and at 134 feet (just over 40m meters) was for a long time the highest dam in Europe. Some others in Spain were also of very considerable size, and it has been justly said that this region was "the birthplace of modern dam building."[11]

The economic prosperity of Spain was not long sustained, and experience gained there was used more fully in France. Several French engineers visited Spain specifically to study its dams. One result, noted briefly in chapter 6, was a reservoir designed to feed water to the great Languedoc Canal, built between 1662 and 1681. The engineering works associated with sixteenth-century agriculture in Italy and the Low Countries were also very extensive, but none was as spectacular as the Alicante dam (figure 17).

More generally, however, the key to agricultural improvement in Europe was not engineering but the balance between livestock and crops. Some crops were now grown especially to feed cattle and sheep, and these animals in turn manured the land. This corrected one major source of imbalance in medieval agriculture, where some regions had specialized too much in rearing sheep, notably parts of England, Spain, and Italy, while elsewhere too few animals were kept and manure was in short supply. The other important aspect of agricultural change in the sixteenth century was the introduction of new crops. Spanish conquests in the Americas had led to the discovery by Europeans of tomatoes, potatoes, and maize. Soon after the conquest of Mexico in 1523, tomatoes were brought from there to Spain. The potato followed around 1570, coming possibly from what is now Colombia, and locally grown potatoes were on sale in Seville during the autumn of 1573.

All these American vegetables were soon adopted in Italy, where the tomato in particular quickly became popular. But new crops were also brought to Italy from the eastern Mediter-

Figure 17
Tibi Dam, near Alicante, Spain, begun in 1580 and completed in 1594. This stone-built dam was designed to provide water for irrigation. Over 33 meters thick at its base, the profile of the dam is tapered and stepped, but is still 20 meters thick at the crest. This makes it heavy enough to function as a gravity dam, though its shape and construction would have allowed a much thinner version to stand as an arched dam. The height, at over 40 meters (134 feet) was not exceeded by any other dam in Europe until the nineteenth century.

(Reproduced by courtesy of Editorial Gustavo Gili, Barcelona.)

ranean, including the wheat from Turkey already mentioned, apricots, and the globe artichoke. Around 1500, the cultivation of rice (already known in Spain) was introduced into Tuscany, and after 1550, into the Po valley and also the hinterland of Venice.

In the latter region, prosperous merchants from the city often initiated agricultural improvements on their country estates, and a map showing the distribution of villas with farm buildings designed by Andrea Palladio also neatly indicates the areas where agricultural development was most actively pursued (figure 18), so closely was he identified with this movement. The villa at Bagnola, for example, stood on the banks of a river whose flooding had recently been brought under control, allowing rice to be grown and the land to be more efficiently farmed. Palladio built large barns here, situated some distance from the villa. In contrast, at Maser, Fanzolo, and other sites (see map), Palladio built villas in which the farm buildings were attached directly to the house in long, projecting wings.

One of the gentlemen-farmers for whom Palladio built a villa was Daniel Barbaro. A scholar as well as a farmer, Barbaro wrote books on mathematics and studied Aristotle and Vitruvius. His house at Maser was notable for the waterworks that supplied fountains in the garden and water for the kitchen. Any overflow or waste from these works was then used to irrigate the gardens. Palladio designed this water system in collaboration with Barbaro, and one is reminded of Alberti's interest in water-supply problems. Water supply was, indeed, one of the branches of engineering most closely connected with architecture.

Another architect with an interest in the subject was Giovanni Battista Aleotti, a contemporary of Palladio who had a reputation as a designer of theaters. He was the architect of several churches and public buildings in Ferrara, a city to the south of Venice and just beyond its borders. As official surveyor there, Aleotti designed fortifications, planned various drainage schemes, and suggested a water-supply system for fountains in the city. But Aleotti's importance rests on his studies of Hero of Alexandria and on the edition of Hero's book on pneumatics he published in 1589. Besides a translation of Hero's text, this contained four appendices dealing with practical devices. One of them showed a four-cylinder force pump driven through a crankshaft by a waterwheel.[12]

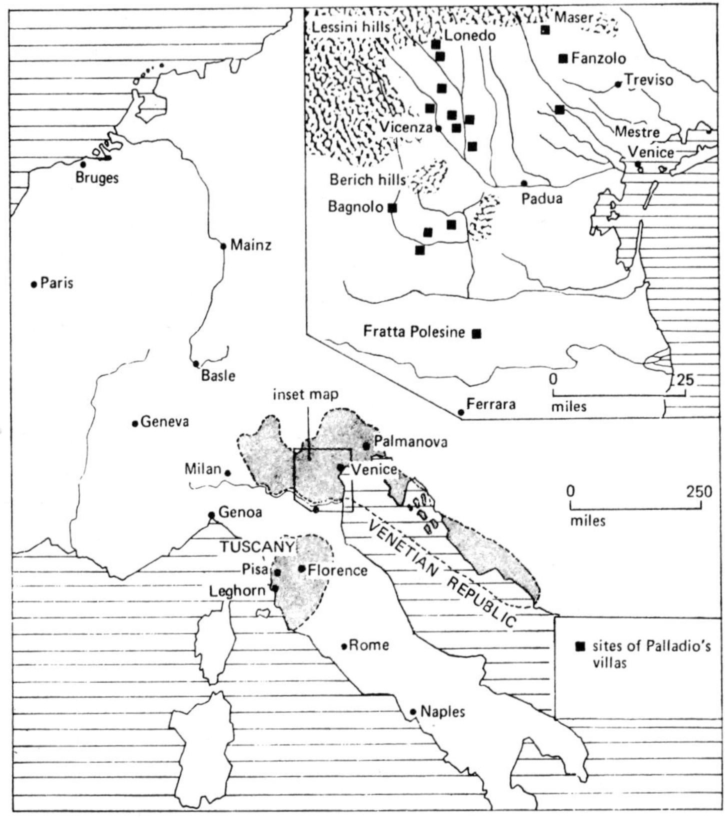

Figure 18
Italy and neighboring areas in the time of Palladio and Galileo. Shaded areas represent the Venetian Republic and Tuscany (the Republic of Florence). The inset map shows the location of villas designed by Palladio in association with land reclamation and other agricultural schemes.

Palladio himself wrote a book, which was published in 1570; it dealt with the design of villas and country houses and was not a work on the theory of architecture.[13] However, the proportions for buildings that Palladio recommended show very clearly that he was using much the same kind of theory as Alberti.

Simple number theories of proportion were no longer so readily accepted as they had been in Alberti's time. Palladio perpetuated the use of such proportions in architecture, but he did not think it appropriate to discuss their theoretical background. At the same time, a new generation of music theorists was becoming skeptical of the Pythagorean number theory of harmony. Among these was Vicenzo Galilei (1533–91), one of several Florentine scholars who were attempting to reconstruct a Greek style of musical drama and whose work was of importance for the origins of opera. As a practicing musician, Galilei had found that number theories were inadequate to explain the complex harmonies used by contemporary composers:[14] and he did experiments with vibrating strings to try to arrive at a better understanding of their behavior.[15]

Galilei's work on this subject was taken further by his more famous son, Galileo Galilei (1564–1642), who wrote extensively on many aspects of mechanics and the practical arts. The style and effect of his writings is, in fact, well illustrated by his comments on the theory of music. Continuing with his father's argument, Galileo pointed out that whether somebody listening to music was conscious of harmony or discord depended on the kind of vibration transmitted through the air to the listener's ear. If several notes heard together produced a regular pattern of vibrations, the eardrum could easily respond, and the listener experienced a pleasant sensation that he called harmony. But if the pattern of vibrations was very irregular, the eardrum would have difficulty in following the motion; the listener would feel discomfort and say he had heard a discord.

The old theory, in which simple ratios were associated with harmonious combinations of notes was a misleading half-truth, Galileo implied. One could not develop a theory of music by tinkering with numbers and ratios, but one should try to understand the mechanics of the vibrating string, taking account of its *weight* and *tension* as well as its length.

Galileo discussed this topic in a book that was largely written by 1610, but published only in 1638. This also covered other types of vibration and discussed how the *weights* of beams and pillars should influence the design of building structures. By implication, therefore, although he did not make the argument explicit, he was challenging the whole range of concepts that linked musical harmony to architectural proportion. Had Galileo been so inclined, he could have gone on to propose a new aesthetic for both architecture and music. But he was content to leave that aspect alone, perhaps because he thought that aesthetics was an unimportant matter of "secondary qualities." Enjoyment of music was no longer imagined to be some kind of response to the harmony of the universe; it was now just a matter of "a tickling of the eardrum."[16]

The theories of music and proportion which Galileo was criticizing had grown out of a strong conviction about the fundamental role of mathematics in nature. Galileo shared that conviction but adopted a different policy for exploring its implications. The Florentine tradition from the time of Alberti had been based on a Pythagorean or neo-Platonic policy; this assumed that numbers and ratios were by themselves the key. Galileo was looking to a different and more practical author from the Graeco-Roman past, and can be said to have adopted an Archimedian policy for investigating the mathematical basis of nature. This focused on *weights* rather than pure numbers and was developed through the study of mechanics, which is the science of supporting and moving weights.

Galileo in the Venetian Republic

Galileo was a Florentine by birth and education, but he spent his most fruitful years at the University of Padua in the Venetian Republic. He was a teacher there from 1592 until 1610, and so was able to witness the last of the big land-reclamation schemes in the hinterland of Venice. His interest in this work is shown by his writings on machines and by the patent he was granted for an unspecified "device for raising water and for most easily watering the land, at small expense and with great utility."[17] This was a pump of some kind, driven by the power of a horse, and it apparently worked well when tested.

Later in life, when he had moved back to Florence, Galileo became involved in other drainage schemes there. He was appointed "superintendent of the waters," and in 1630 drew up a report on the River Bisenzio with the assistance of two practicing engineers.

Galileo's teaching at Padua encompassed the design of fortifications and many other applications of mathematics, such as calculation of the range of a gun. One subject that occupied his attention a good deal, however, was the theory of machines. Here, Galileo was reacting against an overenthusiastic view of the capabilities of machines that had come into prominence through a number of printed books on mechanical invention. These books belong to a tradition that also includes the famous notebooks of Leonardo da Vinci (1452–1519).

Leonardo's work was not published in the sixteenth century and few people knew of it, so it cannot be said to have stimulated other writers and illustrators of mechanical inventions. It was, however, the most imaginative and brilliant work of its kind, exploring ideas for flying machines, paddle-boats and diving suits, as well as recording many existing devices in current use.

A more typical and far more influential piece of work in this same tradition was a book published in 1588 by Agostino Ramelli, a military engineer working in France.[18] Among its nearly two hundred illustrations of machines, more than half showed pumps. Many of these drawings have a certain science-fiction quality. The machines Ramelli drew were speculative inventions, the product of a lively imagination. At times he was led into absurd elaborations of every conceivable kind of pump. Some had curved cylinders and piston rods, while others anticipated the rotary pumps that were to become practicable only in the nineteenth century.

A much more practical book was the "Theater" of machines compiled by Vittorio Zonca of Padua and published in 1607, four years after its author's death. Of all the contemporary works on this subject, Zonca's is said to have been "closest to the actual mechanical practice of the time." Thus, where Ramelli showed elaborate combinations of gears and pulleys in which the effects of friction would more than cancel any mechanical advantage, Zonca usually showed the simplest arrangement of gears that would do the job. And whereas Ramelli largely neglected the industrial arts, Zonca treated them much more fully, giving especially valuable pictures of the silk-throwing mills used in northern Italy, an example of which appears as figure 7. Significantly, the only earlier illustration of these machines with any clarity was one by Leonardo. But even Zonca's enthusiasm could run away with him, for the book ended with a perpetual-motion machine "as fantastic as any."[19]

The element of fantasy in all these books was immensely significant for the future of technology. It reflects an idealization of the machine and an ability to think imaginatively about machines, which was rare (but not entirely lacking[20]) in other civilizations. But this tendency to idealize machines would probably have borne little fruit had it not been complemented by a more sober and utilitarian approach, which is why Galileo is so important. He saw that in most of the current books about machines, one matter badly needed clarification: there could be no question of using machines to overcome nature. Nor would they give one something for nothing. A lever did not create force, even when it allowed one to raise a weight greater than one could ever lift unaided.

In the example shown (figure 19), the short arm of the lever is lifting one end of a block of stone, equivalent to a weight *R*, because somebody is pressing down on the long arm of the lever with a force *E*. Although *E* is a smaller force than *R*, it can lift a block of stone through a small distance but only because it moves through a much larger distance. In other words, the most fundamental advantage of a machine such as a lever (or pulley, or set of gears), is that it allows a small force moving fast over a long distance to move heavy weights more slowly over shorter

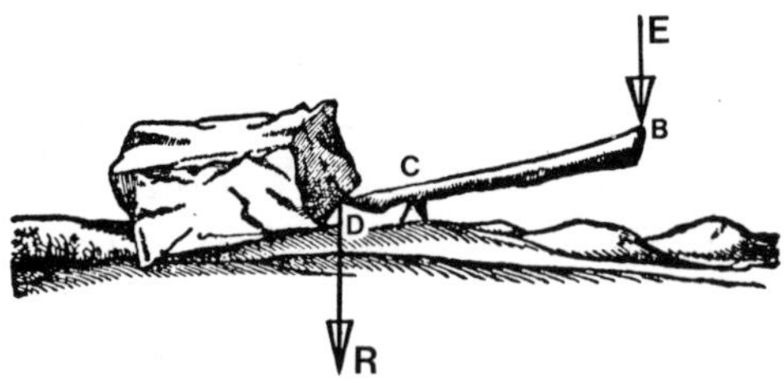

Figure 19
A simple lever supporting part of the weight of a large block of stone, represented by the force *R*. This is balanced by the force *E*, exerted by the person using the lever to lift the stone.

It is possible for force *E* to raise the stone, even though it is smaller than *R*, but only because it moves through a larger distance than *R*.

(Adapted from a diagram in Galileo's *Discorsi e Dimostrazioni Matematiche,* translated by Henry Crew and Alfonso de Salvio as *Dialogues Concerning Two New Sciences,* New York: Dover Publications Inc., c.1950. Reproduced by courtesy of Dover Publications Inc.)

distances.[21] Thus the gain in force is offset by the loss of speed, and hence the machine does not give one something for nothing.

Galileo also noted that some machines offered an advantage of a different kind. They allowed one to harness cheaper sources of power than would otherwise be available: "the fall of a river costs little or nothing." The maintenance of a horse "is far less expensive" than it would be to pay men to do the same work.[22] Thus Galileo ranked economy as "the greatest advantage" of machines, and his account implied the possibility of evaluating the cost of operating waterwheels and horse gins. Galileo lived in close contact with the merchant communities of Florence and Venice. He would know something of the manufacturing enterprises they promoted and the methods they used to control costs and so safeguard their profits.[23] It is a matter of some significance that this kind of cost argument came to the fore in Galileo's formal writings on machines–a subject that could otherwise be treated entirely in terms of academic mechanics.

Before Galileo left the Venetian Republic in 1610, he tackled another problem that had been confusing local artisans and machine-builders. Some of these people had made great claims for machines they had invented and had demonstrated their ideas by constructing models. But there were often disappointments, for it was found that full-sized machines made on the same pattern as the models were often not strong enough to perform their function, or failed in other ways.

More experienced craftsmen knew that there was a fundamental difference between models and fully scaled-up versions of the same design, but could not entirely explain the reason for it. It seemed paradoxical, because the theories of proportion used by architects encouraged people to think only about ratios of lengths and geometrical properties, and these would be the same both for a full-sized object and a scale model. One engineer who commented on this problem was an expert on fortification named Buonaiuto Lorini. After commenting on how people were misled by "the ease with which little models work,"[24] he pointed out that the key factor they neglected was the weight of the materials used in construction.

Galileo had taken up this same point and developed related ideas about scale and proportion while he was working in the environs of Venice. In a shipyard there, he observed that large ships under construction on a slipway needed to be supported by scaffolding although small ones did not. Large ships were weaker in proportion to their size than small ones, and Galileo recognized that the same problem occurred here as with the model machines that did not work on a large scale. Again, the crux of the problem was the weight of the timbers from which either a ship or a machine was made. If a large ship was ten times as long as a small boat, and if all other linear dimensions were ten times as great, then the volume of timber used in the large ship would be greater than that for the small boat by a factor of ten cubed (10^3). Thus the ship would be a thousand times as heavy as the boat, and Galileo saw that much of its strength would be taken up in merely supporting its own weight.

When Galileo turned to a mathematical analysis of this subject, he considered not ships or machines, but a much simpler kind of structure—a wooden beam projecting horizontally from a wall in which one end was firmly embedded (figure 20). Following the traditional approach that emphasized proportion and geometry, one might simply consider the shape of the beam—the ratio of its breadth to its length, for example. But Galileo also discussed its weight (marked *W* on the diagram) and a load (*E*) supported at the end of the beam.[25]

What was novel in his approach was that he saw how the beam could be treated as a bent lever by which a force *R*, representing the longitudinal strength of the beam, supported the two weights *W* and *E*. Then considering a series of progressively larger beams of the same shape, Galileo showed that the weight of the beam increased more than the force *R*, so that eventually there was a limiting size, where a beam could only just support its own weight and could carry no load at all.

Once he had pictured a beam sticking out of a wall in this way, it was perhaps quite obvious that the beam could be thought of as one arm of a lever, so that to complete the analysis it was necessary only to imagine the other arm, which represented the beam's resistance to breaking. What is of special interest about this approach is the way in which Galileo was widening the scope of the theory of mechanics. He could see that levers and loaded beams were both particular cases that could be understood in terms of the same general theory. In principle the theory could be extended to cover the working of every type of machine—waterwheels, horse gins, cranes and so on—and the equilibrium of every type of structure—ships, buildings, animals' skeletons, or the framing of machines. This theory therefore had for Galileo the same very wide significance as harmony and pro-

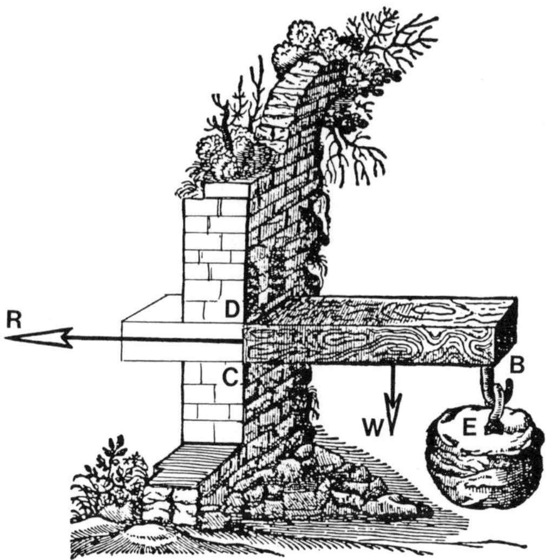

Figure 20
A simplified version of Galileo's theory of the strengths of beams. The beam is firmly embedded in the wall of a ruined building. The weight of the projecting part is *W*, and it supports a further weight *E*. If these weights are large enough to break the beam, it will begin to tear apart at point *D*, and Galileo imagined that the collapsing beam would pivot about *C*. The beam can then be thought of as a lever, analogous to the lever in figure 19, whose pivot is also marked *C*.

If the beam does not break but successfully supports the loads *W* and *E*, there is a resistance to breaking that Galileo imagined acting like a large force *R*. This acts on the short arm of the lever to balance the forces *W* and *E* on the long arm.

(Adapted from a diagram in Galileo's *Discorsi e Dimostrazioni Matematiche*, 1638, by courtesy of Dover Publications Inc., New York.)

portion had for Alberti. It was the one branch of mathematics that provided real clues to the mathematical structure of nature.

Galileo must have been aware that he was overturning the old theory of harmony and proportion when he linked his comments on music to his discussion of scale factors and the strengths of beams. So both the idealistic tendencies in technology that have been described in this chapter, mathematical harmony and mechanical fantasy, were brought rudely down to earth by Galileo's utilitarian outlook.

Galileo considered that the value of machines was to be judged chiefly by their economic usefulness, and he seems to have thought about the economics of building structures also. Analyzing the projecting beam illustrated in figure 20, he noted that it would break at the point where it entered the wall. Toward the end furthest from the wall, the beam would not need to be so strong. Thus material could be shaved off the top of the beam at this end without weakening it, and Galileo calculated exactly how much the beam could safely be reduced in depth. The result, he found, was that the beam would have a curved profile, and the exact form of the curve would be an arc of a parabola. The effect would be to diminish the weight of a beam by one-third without diminishing its strength, "a fact of no small utility in the construction of large vessels, and especially in supporting the decks, since in such structures lightness is of prime importance."[26] Thus Galileo had defined an ideal shape for a cantilever beam, but his concern was with economy, not aesthetics.

Engineering Practice and Architecture

The new thinking about engineering and building with which this chapter has been concerned was the special interest of a small intellectual elite, mainly in northern Italy. Thus it may be felt that the chapter has ignored the day-to-day practice of technology, and that the work of ordinary builders, stonemasons, millwrights, and shipbuilders was perhaps quite untouched by these developments. Although the latter point is partly true, we should not forget that the invention of printing had radically changed the rate at which ideas could spread. Books on machines such as Ramelli's, or on architecture such as Palladio's, seem to have been popular and to have had wide influence.

One result was that a fashion for some aspects of Renaissance-style architecture spread through much of Europe during the sixteenth century, compelling architects to adopt the new scale-drawing techniques and stimulating them to rethink the mathematics of building. Notable examples are Philibert de l'Orme in France and Robert Smythson in England, both of whom came from families of stonemasons and not from the educated elite. In a book he published in 1567, Philibert de l'Orme insisted that every mason on a building site needed better knowledge of geometry and arithmetic if he were to understand the plans drawn up by architects.[27] His own drawing techniques were an interesting synthesis of medieval geometrical methods such as those described in chapter 2 and the new techniques based on measurement. A little later, Smythson was producing drawings of high quality wholly in the new idiom, often with a scale of feet clearly marked on them.[28] Shipwrights were also using scale drawings at this time, long before engineers adopted the technique, one example being the Englishman, Matthew Baker, who made precise drawings of ships' hulls about 1586, sometimes again marked with a scale of feet like Smythson's.

It will be recalled that drawing to scale was linked in its origins to the invention of drawing in perspective. The latter was important for its use in illustrated books on technology, including works full of speculative ideas such as that by Ramelli, and more strictly factual books of which the most famous was by the German author Georgius Agricola. This contained many woodcut illustrations in perspective of machines used in mines (see chapter 5). Innovation and design in engineering depend partly on nonverbal thinking and the ability to manipulate visual images. We may estimate, therefore, that there was a significant stimulus to mechanical innovation from the increased circulation of pictures of machines made possible by printing and the more effective style of illustration made possible by perspective.

As to the ideas introduced by Galileo, which were based on abstract rather than visual thought processes, they probably had little influence on the everyday practice of technology for a very long time. However, his theory of the strength of beams bore fruit in France during the late seventeenth and eighteenth centuries, stimulating the formulation of more sophisticated theories of engineering. It also bore fruit in England, though not until the late 1790s. Then, as we shall see in chapter 7, when iron construction was first used in steam engines, the scaling of parts and the parabolic profiles of beams and connecting rods was due to the direct application of Galileo's thinking.[29]

4

The Practical Arts and the Scientific Revolution

New Kinds of Science and the Study of Pneumatics

Galileo was a major figure in what has become known as "the scientific revolution"—the reassessment of traditional ideas about the natural world that took place in the sixteenth and seventeenth centuries, and the establishment of something recognizably like modern science during the seventeenth century. It was a revolution that most notably changed accepted views about the shape of the universe, with Copernicus in 1543 suggesting that the earth was not fixed in space but moving in an orbit about the sun; with Galileo in 1610 using his telescope to prove that the planet Venus, at least, had such an orbit, and that the moon was a world of mountains and craters; and with Newton in 1687 formulating laws of motion that explained the earth's orbit.

The new science of this time was not only concerned with the motion of the planets, though; it was also concerned with how this motion was related to such things as the trajectories of cannon balls. It dealt with the earth's atmosphere as well as the earth's motion in space; and it encompassed Galileo's theories about machines and the strengths of beams. What technological application was found for all this?

Around 1600, there were already some examples of the practical application of what might be called "science," most notably among the groups of artisans that included opticians, clock makers, and the makers of instruments for use in navigation. Galileo pioneered the use of the telescope in astronomy after its invention by craftsmen of this sort, and he also had ideas about the improvement of clocks. If the motion of a clock was regulated by a pendulum, he thought, instead of by the traditional verge-and-foliot mechanism, the clock would keep time much more precisely. But although Galileo devoted a great deal of effort to the theory of pendulums, he did not persevere with its practical application. So it was not until the 1650s that the first successful pendulum clocks were designed and built, by two men working in Holland—Christian Huygens, a mathematician, and Samuel Coster, a clockmaker. A similar collaboration of "scientist" and clockmaker was that between Robert Hooke and Thomas Tompion in England. Hooke was especially interested in the idea of a balance wheel moving against a spiral spring as an alternative to the pendulum.

Apart from this kind of interaction between "science" and the practical arts, which was of considerable importance at this time, there were other kinds of academic learning which influenced "technology," including the Renaissance interest in Greek and Roman technical literature. To a remarkable degree, indeed, the new sciences of the scientific revolution were built on foundations laid by the ancient Greeks. The critical reappraisal of what was found in Greek writings, and the use of what seemed most valuable, was an enormous stimulus. Galileo's debt to Archimedes is a good example.

The practical arts also benefited from these studies, and the series of discoveries and innovations which led eventually to the invention of the atmospheric steam engine began with a typical late-Renaissance interest in the works of Hero of Alexandria, who described a number of pneumatic gadgets and machines in the first century A.D. The steam engine, which was first successfully operated in 1712, is rated by some as the greatest of the technological fruits of the scientific revolution, but the earliest work relevant to its invention derived from Hero, and seems almost trivial. This work was done by three architect-engineers in the Renaissance tradition who became interested in Hero's writings. They were G. B. Aleotti, surveyor at Ferrara in the 1580s (chapter 3); the architect Giovanni Branca (1571–1640), and Salomon de Caus (1576–1626), a French landscape gardener and deviser of fountains.

Hero's gadgets had a strong appeal for these men, as also for more speculative thinkers such as G. B. della Porta (1535–1615), the author of a book on "natural magic." Porta later wrote "three books on pneumatics" in which there were several ideas for devices modeled on Hero's. Some of these could draw liquids through pipes, or use steam to lift water (see fig. 21). Influenced both by this work and by Hero of Alexandria, Giovanni Branca,

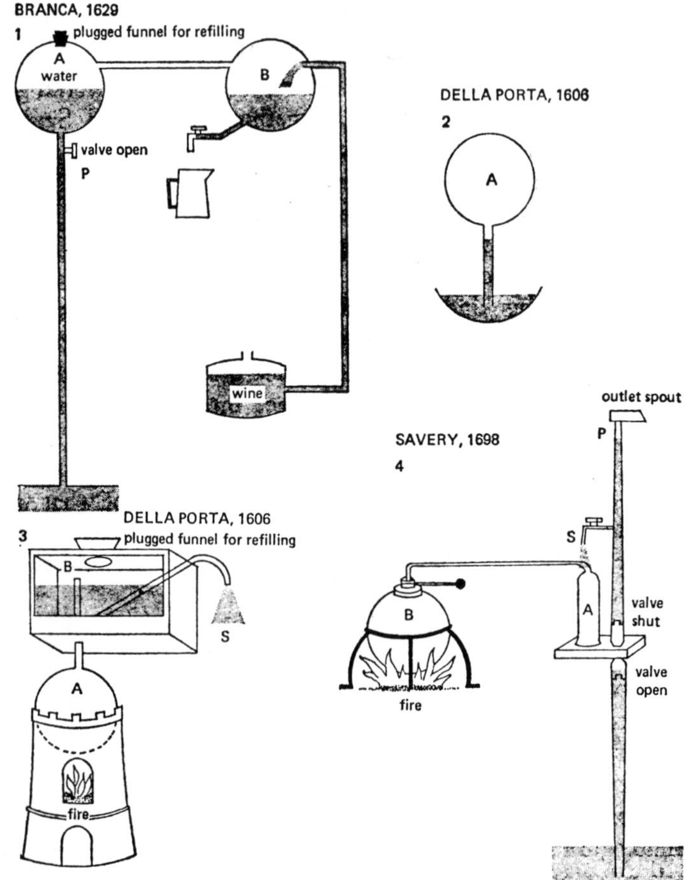

Figure 21
Pneumatic and steam devices owing their origin to late-Renaissance interest in the work of Hero of Alexandria.

1. Design by Branca, 1629. Opening valve *P* releases water from *A*. If the plug is airtight, the air pressure in *A* and *B* is reduced, and wine is drawn into *B*.

2. Della Porta, 1606. Flask *A*, first with its neck upwards, contains a little water that is brought to the boil. When *A* is full of steam, it is inverted as shown. As it cools down, steam is condensed and water is drawn up.

3. Della Porta, 1606. Steam pressure in flask *A* and space *B* forces water up through *S*. The filler funnel is kept sealed.

4. Savery, 1698. The container *A* can be cooled by the spray *S*, and is connected via a cavity in its flat base to the long vertical pipe. With the valves as shown, condensation of steam in *A* by the spray causes water to rise up the vertical pipe to fill *A*. Then if the setting of the valves is reversed and the spray is turned off, high-pressure steam from *B* can force water to the top of pipe *P*.

in 1629, described devices for using steam as a source of power, and elaborated on a wine-lifting gadget invented by Porta.[1]

Salomon de Caus also wrote a book that described pneumatic and steam-powered gadgets, but of greater interest is that in 1609 or 1610 he visited England. There he built pumps and fountains at Hatfield House, at Greenwich Palace and at a house belonging to the Prince of Wales at Richmond. He spent only a short time in England, leaving in 1613, but he seems to have communicated his interest in steam-driven toys to a certain David Ramsay, a member of the Prince's household. Ramsay, like de Caus, was something of a speculative inventor, and took out several patents when this became possible in England after 1624; one of them, dated 1631, was "To Raise Water from Lowe Pitts by Fire."

No details are available to show how Ramsay's machine would have worked, but he does seem to have made the transition from thinking about toys and steam-powered gadgets to considering the application of steam to an urgent practical problem, the drainage of mines. So interest in making a steam engine was aroused in Britain by de Caus, and other inventions followed Ramsay's in 1663, 1682, and 1698. The last date refers to Thomas Savery's patent for a device that first sucked water into a closed container by the condensation of steam and then forced it upward through another pipe by steam pressure. There were many efforts during the seventeenth century to build a practical engine utilizing della Porta's ideas,[2] and Savery's engine can be accounted for simply in these terms.

The Discovery of Atmospheric Pressure

Hero's pneumatic devices were mostly conceived as table-top models, but the early seventeenth-century investigators at first thought that they could be made to any size one liked. It was soon found, however, that siphons, suction pumps, and gadgets like the wine-lift would not work through heights greater than about 30 feet (9 meters). This suggested to Gasparo Berti (c. 1600–1642), a mathematician living in Rome, that one should investigate what happened above this limiting height. He therefore arranged a vertical lead tube about 36 feet (11 meters) long so that it could be filled with water from the top and then sealed. A valve at the bottom was closed while this was going on, and the foot of the tube was immersed in a vessel of water. If, after the top was sealed, the valve at the bottom was opened, some water would be seen to run out. The water in the tube settled with its surface at the 30 foot (9 meter) level, and a space was left above it.

Berti did this experiment in 1641 or 1642, and his ideas were developed in the next few years, first by Galileo's friend Torricelli and then by the Frenchman Blaise Pascal. Between them, these two experimenters established the idea that because air has weight, the atmosphere must exert a pressure. The height of the atmosphere was itself limited, they realized. The air extended only a few miles above the earth. There was a corresponding limit to the height of water in a vertical pipe that the atmosphere's pressure could support, or to the height of a mercury column in one of Torricelli's newly invented barometers. It seemed, in fact, that the weight of air in a hypothetical tube long enough to extend vertically to the top of the atmosphere was equal to the weight of water in a tube of the same diameter but only 30 feet (9 meters) long.

This work, which is described in all standard histories of science, was paralleled in Germany by experiments performed by Otto von Guericke[3] around 1650. Surprisingly, von Guericke seems at first to have had no knowledge of what had been done by Berti, Torricelli, or Pascal, but worked quite independently, probably on the basis of what he had read of della Porta and similar authors. His approach was therefore rather different from that of the Italians. For example, in an attempt to determine whether a vacuum could be created artificially, he successfully pumped water out of an airtight copper vessel. Improving his apparatus, he soon recognized that air itself was a fluid and, like water, could be pumped out of the vessel. It was then a small step to redesign a conventional pump for the purpose, and so he invented an air pump.

A man of practical bent and wide education, von Guericke had helped in the defense of Magdeburg when it was besieged during the Thirty Years War and had worked as a military engineer. It was after the war that he began his work on pneumatics, and this quickly attracted attention. In 1654, when the rulers of the various German states were meeting at Ratisbon, von Guericke gave them a public demonstration of some of his experiments. In one of them, which became famous throughout Europe, teams of horses failed to separate two close-fitting brass hemispheres inside which a vacuum had been produced.

Following this demonstration a man named Caspar Schott was asked to write an account of it. Schott had been a pupil of a priest named Athanasius Kircher, who had actually been present at Berti's experiments in Rome. Indeed, it is possible that von Guericke first learned of the Italian work through Schott. Both Schott and Kircher were Jesuits, and we may comment in passing that they were just two of many seventeenth-century Jesuits who had an interest in science—several in Italy during Galileo's time were expert astronomers. And though these men were often out of sympathy with the new ideas of the time, they contributed greatly to the scientific revolution through their researches and by helping to spread essential information.

Caspar Schott's first account of Guericke's 'hydraulico-pneumatic' experiments was published in 1657 and attracted wide attention throughout Europe. In England, Robert Boyle read the book and immediately arranged for Hooke to build him an air pump like von Guericke's. Meanwhile, Guericke's work was continuing, and in 1661 he did what was perhaps his most important experiment. A cylinder such as might be used as part of a large pump was placed in a vertical position with its open end upward. A close-fitting piston was held against stops at the top of the cylinder by a cord passing over two pulleys. At the end of the cord, several large weights were suspended (see figure 22). Air was then pumped out of the cylinder through a valve at the bottom, and when a vacuum had been produced, the pressure due to the atmosphere was sufficient to push the piston downward into the cylinder, so lifting the weights. The importance of this experiment was that it suggested ways in which the new discoveries about the atmosphere could be given a practical application. It was described in 1664 in Schott's book *Technica Curiosa* and again in 1672 when von Guericke himself wrote a book.

Galileo's Science in France

The cylinder-and-weights experiment of von Guericke had its most fruitful influence in Paris, where an academy of sciences was founded in 1666. Two of the leading members were the Abbé Edmé Mariotte and the Dutch mathematician Christian Huygens. Besides taking an interest in the kind of work von

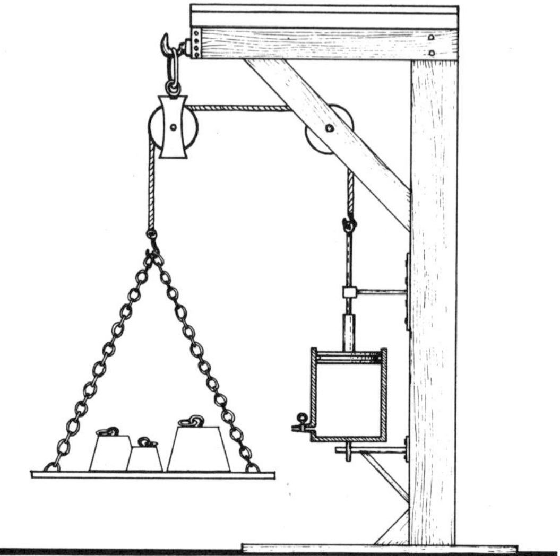

Figure 22
The experiment carried out by Otto von Guericke in 1661. The air has been pumped out of the cylinder on the right. The pressure of the atmosphere on the piston supports, and may lift, the weights at left.

Guericke was doing, Mariotte and Huygens took up many of the problems Galileo and his friends had raised.

Their action was fortunate, because mechanics and practical sciences were not flourishing in Italy, and the inauguration of the Paris academy of sciences almost coincided with the closure, in 1667, of a similar academy in Florence. This largely marked the end of the "Galileo tradition" in Galileo's own former home town. Although the closure of the academy after only ten years may have owed something to the disapproval with which the Church regarded him, even after his death, it was probably also related to the economic decline being experienced by many of the small Italian states. Even in 1610, when he left the Venetian Republic, he was probably aware that having lost its merchant fleets, Venice was now losing its cloth trade.[4] Other states were affected by similar economic problems and by epidemics that brought a high mortality. There was little to attract engineers and inventors, or to provide the technological stimulus that Galileo had enjoyed during his time at Padua.

Things were different in Paris, where Huygens and Mariotte in the 1670s found themselves in an environment in which public works and architectural projects were being put in hand on a large scale. And while Huygens's relationship with the science of Galileo may be demonstrated mainly by his work on the theory of pendulums, Mariotte's had more to do with engineering projects of the kind that had been so prominent in Italy during Galileo's time and earlier. Mariotte had a long-standing interest in hydraulics, which found application in work that Louis XIV had begun at the Palace of Versailles. There was a poor water supply at the site of the palace, but the architects had called for numerous fountains in the gardens, and a quite extraordinary pumping station was constructed to supply them. It was known as the Marly Machine and consisted of fourteen waterwheels that pumped water from the River Seine through a total height of about 530 feet (160 meters).

Mariotte's connection with this undertaking began when he was asked to calculate the best dimensions for the cast-iron pipes that were to be used. This meant calculating the strengths of pipes of various dimensions, and in doing so, Mariotte discovered and partly corrected a defect in Galileo's theory of the strengths of beams.

When he died in 1684, Mariotte left an unfinished book on subjects related to this work,[5] which was published in 1686. Besides material on the strengths of beams and pipes and on hydrostatic pressure, this book contained a theoretical discussion of undershot waterwheels of the type used at Marly. He adapted ideas previously put forward by Torricelli to prove that the force of water on a stationary waterwheel must be as the square of the speed of the current. This calculation was by itself of little value; but in 1704 Antoine Parent, onetime pupil of the editor of Mariotte's work, used it in a more complete theory of the working of waterwheels.[6]

Parent's theory was a remarkable achievement, involving what was probably the first application of the differential calculus to an engineering problem and resulting in perhaps the earliest statement of the mechanical efficiency of a machine. He said that if an undershot waterwheel driving pumps (like those at Marly) was made to pump water back upstream into the source of its own supply, it would pump back only four twenty-sevenths (15 percent) of the water needed to drive it: in modern parlance, its efficiency would be 15 percent—although in fact this figure was too low because of a fault in the reasoning that Parent took over from Mariotte.

Parent himself was not an engineer but a teacher of mathematics in Paris. The young men who came to him for lessons mostly wanted to learn about the theory of fortification and the military arts, and Parent made it his business to gain some practical experience, by attaching himself to an army unit that was at that time—the early 1690s—operating mainly against the Dutch and on the borders of Germany. His scientific interests therefore grew out of a very similar background to Galileo's and included military problems, architecture, and hydraulics. So it seems appropriate that it was Parent who later provided the first fully satisfactory solution to the problem about the strengths of beams that Galileo had raised.

While Parent was working on these problems,[7] Denis Papin, a Huguenot exile from France, was pursuing other scientific ideas of technological significance. Before he left Paris, Papin had worked as Huygens's assistant and in the 1670s had helped to carry out an experiment that owed a good deal to von Guericke. It will be recalled that von Guericke had arranged a vertical cylinder so that when air was pumped out of it, a piston was pushed downward by atmospheric pressure. This pulled on a rope passing over two pulleys, thereby lifting a weight (figure 22).

Huygens and Papin constructed a similar apparatus in 1673 and arranged for a small charge of gunpowder to be exploded inside it. The idea was that this would force most of the air out of the cylinder through a nonreturn valve—a much quicker and easier process than pumping the air out; then, after the gases left inside the cylinder had cooled down, their pressure would be very low, the piston would be pushed downward, and a large weight would be lifted by a suitable arrangement of ropes and pulleys.

The experiment was successful. A very small charge of gunpowder was exploded in a cylinder, and when it cooled, the piston descended and lifted 1,600 lbs (725 kg) through a height of 5 feet (1.5 meters). The cylinder was just over 1 foot (0.3 meters) in diameter. Here was the basis for an engine—but the idea was not developed further. Instead, Papin thought that a more efficient method of creating a vacuum inside the cylinder might be by the condensation of steam. Accordingly he put a small quantity of water in the bottom of the cylinder and applied a flame to make it boil. As the steam pressure inside the cylinder increased, the piston was forced upward until it reached the top. Then the flame was removed and the cylinder was allowed to cool. A vacuum formed as the steam condensed, and the piston was pushed downward by the pressure of the atmosphere. Papin was able to make his apparatus raise and lower a weight by successively applying and then removing the source of heat from under the cylinder.

The English Steam Engine

In 1704, the year in which Parent published his study of the waterwheel, Papin was settled at Cassel in Germany, from where he wrote to Leibniz about the comparative merits of waterwheels and his steam engines as means of driving pumps.[8] By that time he had tried out other types of steam engine, but although many of his devices worked under experimental conditions, he seems never to have perfected any of them as usable machines. This was left to two Englishmen working in Dartmouth—Thomas Newcomen, an ironmonger and blacksmith, with his partner and assistant, John Cawley (or Calley), a plumber. They took the idea of the cylinder and piston that von Guericke, Huygens, and Papin had tried out; they adopted the idea of creating a vacuum by the condensation of steam, which della Porta, Savery, and Papin had all tried; but to these they added a mechanism of great ingenuity for turning valves on and off in the right sequence, and this automatically controlled the supply of steam and water to their machine. In this way a steam engine utilizing the pressure of the atmosphere became a practical proposition.

The Newcomen engine had a cylinder and piston positioned vertically, as in von Guericke's machine. Steam was supplied from a separate boiler and could be condensed inside the cylinder by turning on a supply of cold water, which sprayed up in a jet from the bottom of the cylinder. When this was done, a vacuum was created, and the piston began to descend as a result of the pressure of the atmosphere, pulling on the end of a pivoted beam above it. Newcomen's engine was designed to pump water from mines such as the ones in Devon and Cornwall near his home, and the purpose of the large pivoted beam was to transform the downward motion of the piston in the cylinder into an upward motion of long vertical rods in the mine shaft that worked pumps at the bottom. When the condensation of steam had finished and the piston had reached the bottom of the cylinder, the pumps had completed one stroke and lifted a certain amount of water from the mine. To start a new stroke, a fresh supply of steam was let into the cylinder. It was not high-pressure steam and could not of itself lift the piston, but once it began to enter the cylinder, the vacuum under the piston was destroyed, and the weight of the pump rods was then sufficient to swing the great beam over and pull the piston back to the top of the cylinder. Then the steam supply was shut off, the condensing water was turned on, and the piston began its descent once again (figure 23).

The first engine of this kind about which anything is known was installed in 1712 at a coal mine near Wolverhampton. It is significant, though, that this was not an experimental engine or prototype, but a perfected machine. It seems probable that trials had been made earlier with a similar engine, perhaps in Cornwall,[9] and that Newcomen and Cawley had been in a position to build a workable engine from as early as 1705. Before that, it is said, ten years were spent in experimentation, the main problem being to find the best way of condensing steam in the cylinder. It seems likely, then, that Newcomen's invention originated about 1695.

The puzzle remains as to how much Newcomen knew about the devices of von Guericke, Huygens, and Papin, which seem to anticipate his engine so closely. While Newcomen was still alive, one man who probably knew him wrote that "the Engine now used was invented by Monsieur Pappein and others, as I find sufficiently described in their writings from which Mr. Newcomen began to make improvements."[10]

This is slight evidence in itself, but it seems very likely that Newcomen did know of Papin's work, and in particular of the experiment on condensation of steam in a cylinder. An account of this was first published in the Leipzig journal *Acta Eruditorum* in 1690. It was written in Latin, but a French translation appeared in 1695 and was reviewed in the *Philosophical Transactions* of the Royal Society of London in 1697. Some historians have felt that these publications were all too obscure for Newcomen in Dartmouth to have ready access to them, and there has been much discussion of whether he was in correspondence with members of the Royal Society in London, particularly with Robert Hooke. The evidence is inconclusive, but here it is worth mentioning that Newcomen was a Baptist by religion, that the local Baptist congregation met in a house he leased in 1707, and

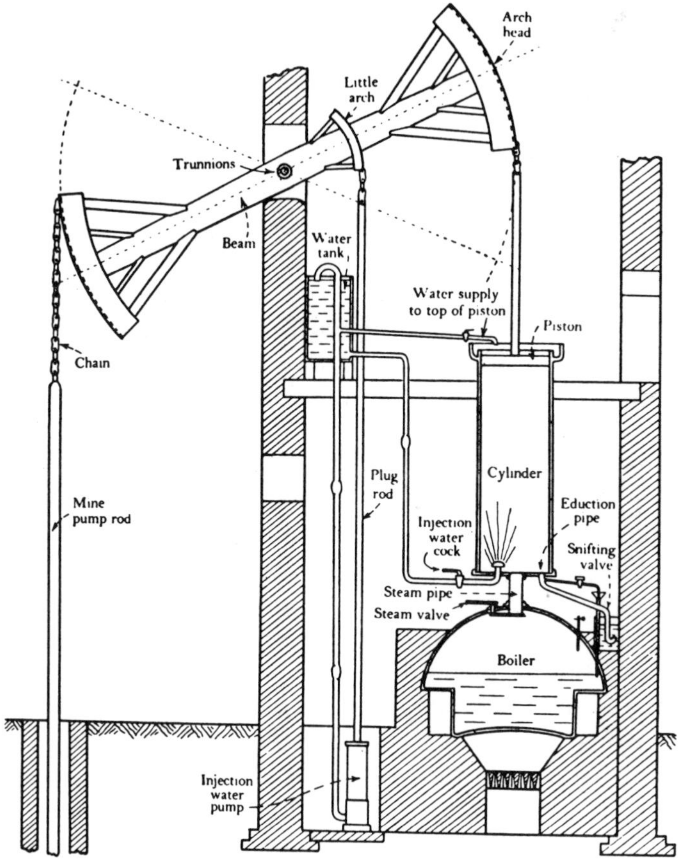

Figure 23
The principle of Newcomen's atmospheric steam engine. The engine is shown at the point where the piston is about to begin its downward stroke. The cylinder is full of steam, but the steam valve is shut and the jet of injection water has just been turned on. This will condense the steam in the cylinder, creating a vacuum, and the piston will be pushed downward by the pressure of the atmosphere. The condensed steam and injection water can drain away through the eduction pipe.

(Source: Figure 7 in H. W. Dickinson, *A Short History of the Steam Engine,* Cambridge University Press, 1939. Reproduced by permission of Cambridge University Press.)

that Newcomen himself was a preacher.[11] From what is known of nonconformist lay preachers at this date, it seems right to assume that most such men, though often self-educated, were also well acquainted with books and knew where to obtain them.

Another point is that Huguenot refugees—that is, Calvinist Protestants from France—had "appeared in large numbers in Devon towns following the Revocation of the Edict of Nantes in 1685,"[12] and some of them settled in the area where Newcomen lived. Thus Newcomen belonged to a cosmopolitan community where it would be easy for him to get translations made of any foreign books that came his way, including Papin's French account of his steam-engine experiments.

Technology and the Scientific Revolution in Summary

Whatever the precise channels of communication by which Newcomen and Cawley heard of the experiments of von Guericke, Huygens, and Papin, Newcomen's "atmospheric" steam engine was closely related to the equipment used in these experiments with respect to both logic and design. It seems that here is a clear case of an important invention deriving from the scientific revolution. But the lack of definite information about a link between Papin and Newcomen has led some writers to think that any such link would have been extremely tenuous, and that the English steam engine was mainly the product of an untutored empiricism. It is pointed out that Newcomen's engine depended on a great deal of purely mechanical ingenuity, and it is argued that "the successful fruition of the heat-engine owes virtually everything but the concept of atmospheric pressure to practical engineers."[13]

The author of this comment goes on in the same passage to minimize the technological significance of Galileo's work and Mariotte's development of it. Their theories were never applied to practical problems, he argues, and were often unrealistic or misleading.

There is an important point here that needs to be considered. The limitations of Papin's steam-engine inventions can be summed up by saying that he took them no further than an exercise in applied science. Similarly, the work of Galileo and Mariotte on the strengths of beams went no further than a scientific commentary on a common experience in engineering.

The social milieu in which all three of these men worked did not challenge them to produce working equipment, but rather to analyze their findings in some form of publication.

Newcomen, by contrast, had regular business with men who built and operated machines for the mines in Cornwall. He had skills as a craftsman and metalworker, but not the inclination or background to spend time writing books. Thus the challenge presented by his social surroundings was to make a machine of practical use in a mining community. In that sense, it is true that Newcomen's engine owed much more to his engineering background than to science.

But the fact remains that this was a period when science and technology were moving closer together, to the benefit of both. Thus Galileo's theories were, *in time,* made use of by British engineers, who were not misled by the oversimplifications in his theory of the beam.[14] And in France, an engineering elite was increasingly trained in mathematics. Parent as well as Mariotte regularly taught young men who later practiced architecture or engineering. One of them was Sebastien Truchet (1657–1729), consultant engineer for the Orleans Canal, who had been a pupil of Mariotte. In another half century there would be schools of engineering in France that took this further. It can be argued that much of the mathematics so taught was yet another manifestation of an idealistic tendency in technology, and that it led to an overemphasis on theory. But although there are instances of this, it must also be said that the highly successful development of engineering in France during the eighteenth century resulted partly from the growing application of mathematical sciences to "architectural" technologies such as hydraulics, bridge building, and fortification. The most important early account of Newcomen's engine to appear in French was contained in an enormous book on canal building, harbor works, and hydraulic machinery that had the significant title *Architecture hydraulique*.[15] Somewhat later, the major developments in technical drawing associated with the name of Gaspard Monge (1746–1818) grew directly out of an interest in the traditional problems of fortification design.[16] Thus the connections between "science" and engineering established during the Renaissance and developed by the scientific revolution did have products of indisputable technological worth.[17] However, the influence of "science" did not always operate in the particular way we expect in the twentieth century, but many ideas passed along the more traditional channel of communication between mathematicians and artist-engineers.

A final blind spot in the conventional historians' approach is the assumption that influence always proceeds from "science" to "technology." Interactions of the opposite kind have often been crucially important. In the seventeenth century, experience with water pumps stimulated "scientific" work on the vacuum and on barometers, and Galileo learned much about mechanics in the shipyards of Venice. The practical arts could also influence "scientific method." The experimental approach in science owed a great deal to people's growing experience of machines and industrial processes. Accurate weighing, as used in the sixteenth century by assayers working at silver mines (chapter 5), became an important part of the method of scientific chemistry when this developed much later.

But the scientific revolution did not just consist of the acquisition of new knowledge and the use of new experimental techniques. It also involved a change in people's outlook on the world—a change in what can be called their "philosophy of nature." On this subject, the men who were responsible for the new knowledge gained at this time held a variety of different views. According to the philosophy of nature accepted by some of them, the universe was in some degree an organic thing, to be studied as one would study the plant and animal kingdoms. Others held that there was something mysterious in nature, which could be understood only by using the concepts and vocabulary of astrology and magic. But the philosophy of nature most strongly associated with the change of outlook that occurred in this period portrayed the universe as being built on the lines of a rather complicated machine. This "mechanical philosophy" demanded a changed outlook because it banished all sense of there being something mysterious or organic about the universe. The new approach emerged only after 1600, largely in the work of Galileo and a few contemporaries, but by the end of the century it was the predominant philosophy of nature among people with scientific interests in Italy, France, and England. It seems worth asking, therefore, whether this mechanical philosophy had any relationship with contemporary machines or with developments in the mechanical arts.

The Mechanical Philosophy

Quite a lot has already been said concerning beliefs about the fundamental importance of mathematics, both as a key to understanding nature and as a guide to the builder or craftsman. After being reformulated by Galileo and others at the beginning of the scientific revolution, these beliefs played a major part in the changed outlook of the time.

But while belief in mathematics and enthusiasm for experiments are rather general attitudes, and may be regarded as among the causes rather than the effects of the scientific revolution, two other ideas may be mentioned that belong more specifically to this period. One is the view that problems may be more easily solved if they are broken down into their component parts and studied piece by piece—this may be called "the method of detail." The other is the notion of a mechanical philosophy, already mentioned, according to which everything in nature, from the orbit of a planet to the functioning of an animal's muscles, operates in some sense like a well-made machine. In 1644 this view was put into words by the French philosopher René Descartes, who said that he did not "recognize any difference between machines that artisans make and the different bodies that nature alone creates."[18]

Galileo never said anything so explicit, but it is worth mentioning that his theory of the strengths of beams was based on a machine theory—the law of the lever—and besides applying this theory to ships and buildings, he also discussed it in relation to animal skeletons, birds' feathers, and the stems of plants—in fact, to all the natural beams and struts out of which these living creatures are built. In doing this, he was one of many seventeenth-century writers to apply mechanical concepts to human and animal bodies. Descartes, however, provides the outstanding example and his writings are of particular interest where they include a description of ornamental water gardens similar to the ones at Versailles for which Mariotte was later to make calculations concerning pipes and waterwheels. Descartes said that the human body was "nothing else but . . . a machine."

Thus you must have seen in the grottoes and fountains which are in the King's gardens, that the pressure which forces up the water from its source is sufficient to move many machines . . . following the complex arrangements of the pipes which carry it. And truly one can compare the nerves . . . to the pipes of these fountains and their machinery; the muscles and tendons to the various other engines and springs . . . the animal spirits to the water which works them; and the heart to the source of water. . . . Further, respiration and other such actions . . . are like the motion of a clock, or rather a mill which turns continuously with the flow of water.[19]

This kind of argument recurred in several of Descartes's writings. Henry Power, a physician practicing near Halifax in Yorkshire, quoted the "strong opinion of Des-Cartes" that animals "indeede are nothing else but engines, or matter sett into a continued and orderly motion."[20] Power was one of several English authors who discussed the mechanistic ideas of Descartes. He had made a series of studies of small animals, using the newly available microscope, and these creatures he described as "Insectile Automata" or "slow-paced Engines of Nature," even though he had some reservations about how far Descartes's mechanistic view could be pushed. The philosopher Thomas Hobbes probably had many fewer such reservations. In the introduction to *Leviathan* he asked, "what is the heart, but a spring; and the nerves but so many strings; and the joints but so many wheels, giving motion to the whole body."

When one considers the motion of a man's arm under the influence of the muscles, or the structural strength of a skeleton, machine concepts can be applied with some exactness. But these examples account for only a small fraction of the functions of animal bodies, and there are relatively few other natural phenomena to which the idea of levers, wheels, and pulleys can be directly applied. Thus when Descartes compared the universe to a watch, he perhaps only meant this to be taken as an analogy. Such an analogy, acceptable in itself, betrays an attitude of mind that is more open to question. The universe is not operated by celestial gear wheels. But in a slightly broader sense, Descartes conceived its behavior to be strictly mechanical in that it could be understood solely in terms of the motions and interactions of solid pieces of matter.

Asked for more details, most of the seventeenth-century thinkers who held such views would have mentioned the motions of the smallest particles of matter that could possibly exist. The ancient Greeks had discussed such particles and had called them "atoms." An account of Greek theories of atoms had been written by the Roman poet Lucretius, and in the seventeenth century this was being widely read after a long period of obscurity. Many people with an interest in philosophies of nature took up the ideas expressed by Lucretius and adapted them to their own purposes. One of the most influential reinterpretations of the atomic hypothesis was by the French priest Gassendi (1592–1655). Galileo and Descartes both talked about small particles, but not in the sense of indivisible atoms.

What attracted all these authors was that Lucretius used his atoms to draw a somewhat mechanistic picture of the universe, in which the world was created by the random motions and combinations of the atoms, and in which the planets were moved in their courses, "on the same principle as we see rivers turn water-wheels."[21]

This ancient work, then, supplied arguments that very neatly supported one of the chief trends in the thought of the scientific revolution. It provided a model of the universe that was not only mechanical in concept but would also justify the current belief in measurement and mathematics as keys to understanding nature. The point was that in visualizing the basic particles of matter, one could think of them solely in terms of measurable quantities such as shape, speed of motion, mass, and position. The many unmeasurable things people experience—for example, warmth, smell, color—could be explained, again in principle, by the motions of small particles and the reactions of the sense organs to them. Thus Galileo thought that he had explained away musical harmony by talking about the reactions of the eardrums to motions in the air; and elsewhere he wrote that "tastes, odors, colors and so forth are no more than mere names so far as pertains to the objects in which they reside, [and that] they have their habitation only in the senses."[22] They were "sensations" and not "real qualities."

Authors who discussed such ideas included Antoine Parent,[23] the teacher of mathematics already mentioned, and in England, Henry Power, Thomas Hobbes, and also John Locke, the philosopher. In their writings, such people sometimes identified their outlook with "the mechanical philosophy," although the more specific term "corpuscular philosophy" was more commonly used.

The Method of Detail

In drawing a distinction between "real qualities," like length and weight, which were measurable, and "sensations," like tastes, odors, and colors, which were to be explained away—a distinction for which Locke later used the terms "primary" and "secondary" qualities—Galileo was doing more than argue for a mathematical science of measurable quantities. He was also analyzing experience and breaking it down into its parts.

It is this analytical habit of mind that, after the example of the nineteenth-century author John Stuart Mill, is termed "the method of detail": a form of thought that operates on several levels. Thus the desire to think of the ultimate particles of matter might be seen as one of its aspects, but its real value and effectiveness was as a means of solving problems. An example has already been briefly mentioned, for when Mariotte was asked to determine the best dimensions for pipes at Versailles, he broke the problem down into several more detailed questions—about how freely the water would flow in pipes of different bore, about the bending strength of a pipe and about its bursting strength when filled with water under pressure. Taken as a whole, the problem might have seemed impossible to answer, but its component parts were reasonably simple.

Similar examples can be drawn from the work of Galileo—as when, faced with a question about the strengths of ships' hulls, he avoided discussion of the problem as a whole by analyzing instead the strength of a single beam. On another occasion he explicitly acknowledged that the new science of his time was achieving results by turning away from many of the big problems that philosophers had traditionally studied and looking instead at the details of the natural world. Taken by themselves, such details might seem trivial or uninteresting, but they did at least present problems that could often be solved, whereas the larger issues were usually baffling. A case in point, thought Galileo, was a question "relating to pendulums, a subject which may appear to many exceedingly arid, especially to those philosophers who are continually occupied with the more profound questions of nature. Nevertheless, the problem is one I do not scorn."[24]

What was involved in this kind of attitude was neatly summed up a hundred years ago by John Stuart Mill in talking about "habits of thought and modes of investigation which are essential to the idea of science." The lack of these before the time of Galileo had made science "a field of interminable discussion leading to no result"; and the essence of the new method introduced at this time "may be shortly described as the method of detail; of treating wholes by separating them into parts . . . and breaking every question into pieces before attempting to solve it."[25]

The application of the method of detail in a practical and technical context can be illustrated by the varied career of Sir William Petty (1623–87), who has already been mentioned in connection with Thomas Hobbes. Petty studied medicine at Leyden in Holland and taught at Oxford; in 1652 he became Physician General to Cromwell's army in Ireland. There, he very efficiently reorganized the army medical service, and at the same time began to take an interest in the country where he was working.

Petty shared the contemporary belief in the value of measurement and mathematics, but applied it in new and original ways. In later life he became a powerful advocate of the need for governments to collect statistics of population and trade, and himself made well-informed estimates of the populations of several countries, including Ireland. Cromwell's military campaigns there had been accompanied not only by terrible destruction, but by outbreaks of plague. Between "23rd October 1641, and the same day, 1652," Petty estimated that the population had fallen from 1,466,000 to 850,000. There were 504,000 Irish and over a hundred thousand English and Scots who "perished and were wasted by the Sword, plague, famine, hardship and banishment."[26] These figures are probably not greatly exaggerated.

The work of Petty's that is of most importance here arose out of an official policy for confiscating Irish lands and using them to pay debts that the English government otherwise had not the resources to meet. To ensure that the award of lands to creditors would be effectively carried out, it was necessary to make a detailed survey of Ireland in a very short time. This job, which seemed almost impossible, was undertaken by Petty with great efficiency, and his map of Ireland, published in 1673, is said to be the most accurate map of any country produced up to that time. But the greatest achievement in making the survey was not so much its accuracy as its organization. Twenty-two counties were to be surveyed in fifteen months (December 1654 to March

1656) by a force of about a thousand men. Many of these were soldiers, unskilled in surveying. The skilled work Petty reserved for his few experienced surveyors, while the other work was done by soldiers, who were given some elementary training for such jobs as taking measurements in the field:

> He divided the whole art of surveying into its several parts, viz. 1, Field work; 2, protracting; 3, casting; 4, reducing; 5, ornaments of the maps; 6, writing fair books; 7, examination of all and every the premisses; withal setting forth, that for the speedier and surer performance he intended to employ particular persons upon each species, according to their respective fitness and qualifications.

Many of the surveyors' instruments were made in a workshop which Petty set up for the purpose; he divided "the art of making instruments, as also of using them, into many parts"—wiremakers made the measuring chains, watchmakers made compasses, and separate workmen were employed for making the wooden and the brass parts of instruments. Finally, a more skilled man checked, adjusted, and calibrated them.

In his essays on "political arithmetic," Petty discussed this method of organizing work. In watchmaking, for example, if "one man shall make the *wheels,* another the *spring,* another shall engrave the *dial-plate,* and another shall make the *cases,* then the *watch* will be better and cheaper, than if the whole work be put upon any one man."[27]

This procedure is an example of the division of labor, the principle of which was given its classic expression a hundred years after Petty, by Adam Smith. As a method of organizing a labor force, it is more usually discussed in connection with the rise of factory industry than with the scientific revolution. But in the work of Petty, a man fully in touch with the science of the seventeenth century and deeply influenced by its mathematical emphasis, one can see that the division of labor, as a concept, sprang from the same source as the method of detail—from the recognition that a job is often done better if broken down into elementary constituent parts.

The Disenchanted Outlook

Although the method of detail can be thought of as a technique of analysis that does not depend on any particular philosophy of nature, it was closely linked historically with mechanical analogies of the world. The mechanical models of natural phenomena discussed in the seventeenth century were always based on imagined assemblies of wheels, levers, or corpuscles. They therefore always encouraged the study of problems in terms of these basic components.

Such ideas seem highly relevant for understanding some of the difficulties surrounding technology and its applications in the twentieth century. The seventeenth-century mechanical philosophy has persisted, in some of its fundamentals, down to our own time, and may have led us into some of our modern dilemmas.

For example, the idea of the natural world as being something mechanical freed people from any old-fashioned doubts about whether, in making a machine or digging a mine, they might be encroaching on the prerogatives of the Creator. The thirteenth-century idea of clocks or cathedrals as symbolizing heavenly things—the idea of their construction as a reaching-out toward the source of light and life—had been replaced by the notion that the heavens themselves were little better than a piece of well-made clockwork. So there was no particular reason for feeling any humility before nature. It was a machine that one could tinker with. And the machine analogy gave no warning that there were checks and balances in nature that could easily be upset, because seventeenth-century machines did not incorporate feedback loops or any other automatic control systems to prevent them getting out of control or running away. The only exception was the escapement mechanism of clocks, but this evidently did not prove sufficiently suggestive.

The mechanical philosophy is sometimes described as a disenchanted view of nature, because it left no room for any mystical appreciation of natural phenomena, and it outlawed astrology and magic. There were immense benefits in terms of clarity of thought and ability to analyze problems, but there were disadvantages too. Links between different aspects of subjects being studied came to be habitually ignored, and objects were isolated from their contexts. Thus it would not be fanciful to see the roots of the twentieth century's environmental crisis in the habits of thought established at this time. For too long, people have found that the best way of solving problems in technology has been to concentrate on details while ignoring wider implications—and that has usually meant ignoring consequences for the environment.

However that may seem from a modern standpoint, we should still recognize its early development as part of an *idealistic trend* in the technology of the seventeenth and eighteenth centuries. This trend can be characterized as an increased tendency to use analytical and "rational" approaches, sometimes in the form of the "method of detail" applied to the analysis of work (as in the division of labor), sometimes in the form of mathematical discussion of subjects like the strength of beams, and sometimes simply through measurement and scale drawing, as discussed in the previous chapter. Although these may seem like a rather miscellaneous collection of new techniques, cumulatively they amounted to a considerable enhancement in the ability to analyze technological problems.

Direct applications of scientific ideas, as in the development of pendulum clocks, or in the use of ideas about atmospheric pressure in steam engines, were relatively infrequent, but the spread of a more rational, analytical, even mechanistic approach was of considerable importance for the *organization* of new industries as well as for the invention of new machines, a point discussed more fully elsewhere.[28]

5

Social Ideals in Technical Change: German Miners and English Puritans, 1450–1650

Introduction

In nearly all accounts of the "scientific revolution," including the one given in the previous chapter, Italians and Frenchmen figure prominently, but, on the whole, Germans are conspicuous by their absence. Otto von Guericke and the astronomer Johannes Kepler are always mentioned by historians, and sometimes also the German Jesuits Kircher and Schott. Yet although von Guericke and Kepler contributed greatly to contemporary understanding of nature, and although Kircher and Schott helped to spread knowledge of recent discoveries, they do not fit easily into the conventional picture of the "scientific revolution." In outlook and culture, these men differed markedly from Galileo, Descartes, and their associates; and the same was true for other Germans, to be mentioned below, whose interests were in chemistry, mining, or metalworking.

One way of describing the approach to "science" adopted by some of the German school is to mention again the different philosophies of nature that were current in the sixteenth and seventeenth centuries. Some people still felt that there was something ultimately mysterious at the heart of nature and were not averse to talking about magic or alchemy. There were hidden "virtues" or powers in material things; numbers and geometrical shapes had mystical significance; and invention or discovery was thought to be the result of inspiration or divine illumination.

Such beliefs derived partly from the neo-Platonic ideas that had for long lain dormant in European thought (chapter 2) but that had been revived and reinterpreted in Florence just before 1500 (chapter 3). Along with other Renaissance ideas, Florentine neo-Platonism had its share of influence in Germany, where current social and religious problems gave it new emphases and applications. Under such influences there developed an approach to "science" and to the practical arts that contrasted sharply with the mechanistic outlook of Galileo and Descartes.

These somewhat mystical attitudes to the study of nature were clearly *not* "scientific" in any modern sense. Historians have therefore tended to exclude them from descriptions of the "scientific revolution," and so to ignore much German work that might otherwise seem relevant to the development of science. The historian of technology cannot so easily dismiss the Germans, however, because from about 1450 they were preeminent in the arts of mining and working metals, and their mystical view of nature played an essential part in their practical and technical achievements.

Metal-producing mines were to be found among the mountains of eastern and southern Germany, and metal-using industries in some of the larger towns. Augsburg and Nuremberg were two of the most advanced centers, and printing, clockmaking, tin-plate manufacture and gun casting flourished there, as well as the more traditional cloth manufactures. Augsburg was situated at the northern end of an important trade route opened up in the fifteenth century. This crossed the Brenner Pass and gave south German merchants easy access to Venice. There they learned the Italian system of banking and finance, and soon were able to establish their own banking houses in Augsburg.

The Practical Arts of the German-Speaking Countries

One difficulty in speaking about Germany in the sixteenth century is the lack of any precise boundaries to which that name referred. Most of central Europe was occupied by German-speaking peoples, and there were German settlements far to the east in Poland and Hungary. But Germany as a political unit did not exist. Most places where the German language was spoken were part of the Holy Roman Empire, but the Empire was a hotchpotch of small independent states like Saxony and self-governing free cities like Augsburg and Nuremberg, with Austria the only "great power" among them. Saxony was important for its silver mines, and copper, lead, and silver were also mined in the Harz Mountains, an area criss-crossed by a bewildering tangle of boundaries defining several small states.

Mining was also important in the Austrian Empire, with many mines in the Tyrol and Bohemia and scattered mining settlements in Hungary, most notably at Schemnitz (figure 24).

In many of these areas, copper, silver, and lead had been worked from a very early date. But there had been a decline during the fourteenth century, owing to the effects of the Black Death, and because the most accessible ores had been worked out. After 1450, new investment and the introduction of new techniques made recovery possible, and the output of the mines greatly increased.

Before this period, most of the ores had been quarried or mined in shallow pits, with few real shafts or tunnels. But now shafts could go deep, with waterwheels or horse gins used to lift material and pump water from the lowest levels. The deepest mines went down to about 600 feet (180 meters) below the surface, with underground workings spreading out horizontally from the shafts. Machinery was applied to crushing the ore, smelting processes were improved, and there was a steady expansion and development up to about A.D. 1530, when something like 100,000 men were employed in mining throughout the Holy Roman Empire.

The rapid development of mining between about 1450 and 1530, and its relative stagnation later in the sixteenth century, must be understood in terms of the metals that were produced and the ways in which they were used. The two most important were silver and copper. Silver was sometimes found "native," but in many places the ores of copper, silver, and lead were found together, and the metals had to be separated in the smelting process. The actual quantity of silver in relation to other metals was usually very small, but its price was high; for many mines, this was the greatest source of revenue.

The importance of silver arose from its use for making coins. Gold and copper were also used in the coinage of all European countries, but silver coins figured most prominently in ordinary commercial transactions. From about 1450, there was a steady growth of commerce throughout Europe, with a corresponding increase in the amount of silver coin in circulation. Thus mines in the Freiberg district of Saxony (see figure 24), which produced on average about 16,000 ounces of silver each year during the fifteenth century, boosted production to an average of nearly 150,000 ounces per annum by 1530 and continued at that level for several decades.[1] Besides silver and copper, a little tin was

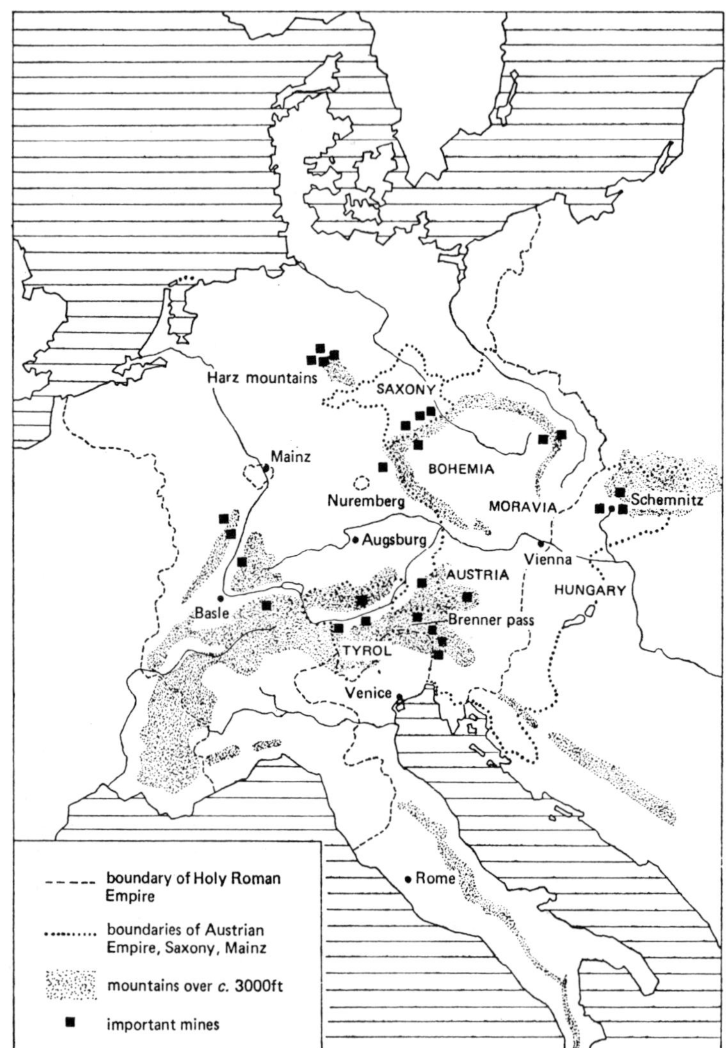

Figure 24
Mining in Europe in the late sixteenth century.

also produced in Saxony and Bohemia, and was used to make tin-plate: that is, sheet iron coated with tin by being dipped into the molten metal. A more important use of tin, though, was in the production of bronze, an alloy consisting of one part tin with about eight parts copper. This was much in demand for making gun barrels, which at this time were either cast in bronze, by means of techniques that had been developed earlier for making church bells, or were fabricated in wrought iron by welding pieces of metal together.

Around the middle of the fifteenth century, guns began to be far more widely used than before, and the resulting demand for large quantities of copper and tin was an important stimulus to the revival of mining at this time. The growing demand for guns probably also provoked thought about using cast iron and led to experiments with that material. Casting iron was more difficult than making similar castings in bronze because of impurities that entered into combination with the molten iron and spoiled its quality. Wrought iron was not subject to this problem, because it was produced in furnaces that were only hot enough to soften the metal, not to melt it. Under these conditions, quality could be more certainly controlled.

However, the quality of the material used to make cannon balls was not too crucial, and they were cast in iron from quite early in the fifteenth century. Blast furnaces for producing cast iron were in operation by 1464, when one was described by Antonio Filarete, an architect from Florence. But only after 1500 did it become possible to cast iron barrels for cannon, and it was not until much later that cast-iron cannon could rival bronze ones in reliability.

Germans were not prominent in the development of cast iron. Early blast furnaces were to be found in Italy, the Low Countries (Liège), and England (Sussex); and by the mid-seventeenth century Sweden was exporting nearly a thousand tons of cast-iron cannon annually. But in bronze cannon manufacture, German expertise was unequaled. In 1592, it was said that German products were the "best available" because the Germans "do things more accurately and more patiently . . . and enjoy a greater and better supply of copper and tin, with which materials they produce excellent bronze. . . . After the castings of Germany, the castings of Venice are considered very good: in Venice the German style and rules are very strictly followed."[2]

There were many other metallurgical arts at which the Germans excelled. For example, there were the arts of assaying and smelting metals, knowledge of which contributed to the early growth of chemistry. Smelting provided empirical experience of chemical reactions; but, more important, assayers had to weigh small quantities of metal with great accuracy, and they had to observe the progress of the metal from one process to another, carefully weighing in at each stage. Similar techniques were later used in chemistry, and the assayers' work probably contributed to their development as well as helping to bring out the relevance of assumptions concerning the conservation of matter.

Bankers and Assayers

It is possible to illustrate several aspects of this wide range of practical arts by looking at the way in which knowledge of German metallurgical techniques was carried across Europe. In many instances, German craftsmen were sent to distant places as agents of the Augsburg banking firms.

During the sixteenth century, bankers such as the Fuggers, the Haugs, and the Welsers were of major international importance; the Fuggers, indeed, probably had more power and influence than any earlier finance house in history. And wherever in Europe an Augsburg financier had an interest, there German craftsmen might be sent to improve techniques and the general efficiency of the operation.

Thus, for example, German technicians were sent to mines in Spain by the bankers of Augsburg and took with them an expertise in pumping machinery, which contributed to the construction of a remarkable but commercially unsuccessful water-pumping station at Toledo. German miners coming to England taught the English the principle of the railroad. Simple types of railroad had been used in the German mines during the sixteenth century for moving small wagons of ore in underground galleries. In England, railroads were used principally on the surface, especially for moving coal from mines near Newcastle to boats on the Tyne. Likewise, it was a German, Burchard Kranich (1515–78), who introduced water-powered ore-crushing machinery at tin mines in Cornwall.

One factor that encouraged German bankers and miners to explore the prospects for mining in places as distant as England was that after about 1550 the mines in Germany were becoming less profitable. Thus the Augsburg bankers possibly hoped, although largely in vain, that richer and more easily worked ores might be found elsewhere in Europe. The cause of the falling profits was that mining in Germany now faced competition from very rich mines that the Spanish had acquired in Central and South America. After the voyages of Columbus in the 1490s and the conquests of Mexico and Peru, completed in 1535, the Spanish were quick to plunder the wealth of the Aztec and Inca civilizations and to discover where silver and gold could be mined. Large shipments of gold and silver began to cross the Atlantic during the 1530s, and by 1600, when this traffic reached a peak, Spain was importing 9,500,000 ounces of silver a year[3]—probably about ten times as much silver as all the mines of Europe were producing annually.

This massive import of bullion had much the same effect as the uncontrolled printing of bank notes would have. Spain experienced a rapid inflation, which influenced other countries and added to an already rising trend in prices and costs; in Saxony, prices rose by about 60 percent between the 1530s and the 1590s. Mining operations in Saxony would become more costly in about the same proportion, while the Spanish imports were keeping the price of silver at a low level.[4] It is not surprising, then, that the bankers of Augsburg were stimulated to look for cheaper sources of silver.

Thus, for example, there was German interest in the ores of copper and lead with traces of silver that existed in the English lake counties, near Keswick. With the help of finance from the Haugs of Augsburg, the Mines Royal Company was set up in 1568 and worked these ores during the last three decades of the sixteenth century. This company had thirteen of its twenty-four shares in English ownership, but the remaining eleven were held by the Haugs through their agent in England, Daniel Hochstetter. Skilled in assaying and smelting techniques, Hochstetter had first to experiment with various methods of refining the local ores in order to discover whether their tiny quantities of silver could be extracted economically. He tried ores from several districts—the Newlands valley, Borrowdale, and Grasmere—and tested several smelting processes and combinations of processes. It was inevitable that some silver would be lost during extraction, and some would remain mixed with the copper and lead that was also produced, but Hochstetter hoped to discover which process gave the best return for the cost of the fuel and labor employed.

In one set of experiments, carried out in 1567, he investigated a process in which an ore or "stone" that had already undergone a preliminary smelting was melted with lead and lead ore. During this operation, the silver was expected to become alloyed with the lead, from which it could be separated fairly easily. Hochstetter used six different mixtures of stone, lead, and lead ore, with "sundry ways tried in melting as we might continue the same with profit." He set out the results of these trials with the quantities of ore and metal expressed in hundredweights (cs.), pounds (lbs.) and half-ounces (lots):[5]

10 cs. stone at 1 lot [of silver per hundredweight]	10 [lots]
3 cs. lead	
4 cs. lead ore	
3 cs. Grasmere glance ore	$4\frac{1}{2}$
	$14\frac{1}{2}$ lots of silver
Out of which was made:	
Lead. 314 lbs. at 2 lots [per cs.]	$6\frac{1}{8}$
Stone. 13 cs. which was dried and brought to 160 lbs. [crude copper] containing 2 lots per cs. [copper]	3
	$9\frac{1}{8}$

The sensitivity of Hochstetter's assays, which could detect one part of silver in 5,000 parts of ore, using quite small samples, is most impressive. But the efficiency of the extraction process seems very poor. Less than half the silver in the "stone" and the ore was concentrated in the lead, and a third was lost altogether. Smelting techniques that worked well with the ores found in Germany probably needed considerable modification before they could be applied to the poorer and rather different English ores.

Hochstetter's comments and figures were written up from his notebook as letters to his backers in Augsburg; it is very clear from the way they are set out, and from such phrases as "3 cs. stone at $1\frac{1}{2}$ lot the cs," that his approach to his calculations was derived from accountancy. One is reminded that Hochstetter himself had come from one of the famous Augsburg merchant families, and so would naturally use the numerical methods familiar from commercial practice. His approach also seems to imply some belief in conservation of matter. If the process were perfected, all the silver in the original ore could be extracted—the apparent loss of silver was simply due to its becoming mixed with slag and other waste materials.

Besides making assays of the ores and their products, Hochstetter studied the cost of smelting in terms of fuel and wages. Thus, for example, one bucket or kibble of the mixed copper and lead ore from Caldbeck would cost, he thought, 11d. to dig from the mine, 3d. for carrying to the smelter, and 14d. to extract the metals. Much of the latter sum was accounted for by the cost of fuel. So the cost of mining and processing 5,000 buckets of ore: that is, about 400 tons, was likely to be about £590. Hochstetter's tests suggested that the yields and revenues from this ore would be:[6]

30 [tons] of lead at 30/-	£ 45
2,187½ oz. of silver at 4/11	528
75 [cs] of copper at 53/4	200
	£773

So there should have been a reasonable profit, mainly because of the high value of the small quantities of silver in the ore. But through no fault of Hochstetter, the Mines Royal Company never sold sufficient metal to cover its costs during the time he was associated with it, and in 1574 its German backers went bankrupt. After that date, activity in the mines slowed down, and only small quantities of metal were produced. Then in 1602 the continuing unsatisfactory situation led to the appointment of a commission to look into the affairs of the company, and in the course of this investigation a whole series of new figures was compiled to show the cost of operating the different mines and smelters. In addition, an attempt was made to assess the running costs of some of the machinery. At one mine there was an "engine"—a waterwheel—that drove pumps and also served for lifting ore and spoil from the bottom of the mine. The cost of running this machine and of pumping water included "reparation above the wheel, pump shafts, bolts, showing of staffs, and like necessaries." There were also payments for "shoose and oil for the gudgeon" and for a "rope for the engine to draw up ore."

Recent surveys of old workings in the Newlands Valley have identified the site of this waterwheel.[7] It was 20 feet (6 meters) in diameter and was located underground on the main level of the mine, at a point where a shaft had been dug down from the level to follow a vein of lead ore. Water to drive the wheel came from a small dam in the next valley about 1,200 yards (1,100 meters) distant. It flowed from there via a wooden trough dug into the hillside and entered the mine via a tunnel known as the "back level."[8] From this there was a considerable fall to the waterwheel, which was probably of overshot type, and from where the water drained out of the mine via the main adit opening into the Newlands Valley.

The Literature of Mining

Technical literature, in the sense of practical manuals and textbooks, scarcely existed before 1450. The introduction of the printing press brought about a radical change in technology simply by making such literature possible. Handbooks on mining and metals began to appear early in the sixteenth century; then, a little later, much more ambitious works on the subject were written, most notably by Georgius Agricola (figure 25).

Agricola was the Latin pen name of George Bauer (1494–1555), a physician who worked among the miners of Saxony and Bohemia. Although a local man, he had studied the classics at university in Leipzig before going to Italy in 1524, where he studied medicine, chiefly at Bologna. He was very receptive to the Renaissance culture of Italy, and like the artist-engineers discussed in the previous chapter, became interested in what Greek and Roman authors had to say about various technical subjects. There was no Roman book on mining to correspond with Vitruvius on architecture. But Agricola read a wide range of Greek and Roman literature, noting every small reference to mining, minerals, and metals. He noted the Latin technical terms that Roman authors used because he wished to write in Latin about current mining technology. In some cases he had to devise new Latin terms for techniques unknown to the Romans. He evidently believed that by writing about mining in a good Latin style he could convince learned men that here was a practical art worthy of their attention. The social status of Italian architects had been enhanced by their activities as scholars and mathematicians. Would not mining benefit in a similar

Figure 25
Laboratory-scale method for separating silver from gold—a typical illustration from Agricola's *De re metallica.* Various items of chemical equipment are illustrated. On the right is seen an assayer's balance such as Daniel Hochstetter must have used.

(See English edition of Georgius Agricola, *De re metallica,* trans. Herber Clark Hoover and Lou Henry Hoover, 1912; new edition, New York: Dover Publications Inc., c.1950, p. 446. Reproduced by courtesy of Dover Publications Inc.)

way if there was a learned literature concerning it? Agricola argued that "agriculture, architecture, and medicine are . . . counted amongst the number of honourable professions . . . mining ought not to be excluded from them."[9]

Agricola's interest in mining had first been aroused when, on completing his medical studies, he began to practice as a physician in the recently founded mining town of Joachimsthal in Bohemia. With Lorenz Bermannus, a local man who had also had a classical education, he began to tackle the problem of bringing the very practical art of the miners into the scope of Renaissance learning. How well he succeeded in this was affirmed by no less a person than Erasmus, who was clearly impressed by the scholarly style Agricola brought to his highly technical subject. Erasmus, in fact, wrote the preface to Agricola's first book on mining, which was published in 1530 by the Froben press in Basle, where many works by Erasmus himself had been issued. This first book took the form of a dialogue in which Bermannus was portrayed interviewing two learned physicians about the mines and minerals of Germany.

In 1533, Agricola moved to Chemnitz in Saxony, where he wrote some half-dozen more books on mining and mineralogy, which appeared from 1544 onward. His masterpiece was *De re metallica libri XII*: that is, "twelve books on metals," which was published in 1556, just after his death.

Agricola's writings were careful and objective. He was a good observer and described assaying and smelting processes with a minimum of "controversial points of theory." One author has said that in this respect "Agricola was the herald of seventeenth century thought."[10] The comment is apt, for in many ways Agricola's outlook and style was typical of the early "scientific revolution."

This point, combined with the evident importance of Agricola for the technical arts and "science" in Germany, helps to underline the paradox of the apparent unfruitfulness of the scientific revolution in Germany. Agricola's approach was less influential than it might have been because for many mining craftsmen his scholarly Latin was merely an obstacle to understanding. German translations were published, but some of them were of poor quality. The only author in the same field who might be compared with Agricola in style was an Italian, the mathematician Vanuccio Biringuccio, who traveled in Germany collecting material for a book on the metallurgical arts. He was influenced by Agricola's first book on mining, the dialogues with Bermannus, and Agricola in turn benefited from reading Biringuccio's *Pirotechnia* (1540).

The Outlook of the Miners

Although German miners in the sixteenth century clearly had a very different outlook from Agricola's, their attitudes are more difficult to document. Most of the information that is readily available was compiled by people who, like Agricola, were not themselves either miners or craftsmen. One account concerning an individual miner that is often quoted, however, deals with a man who worked in the copper mines of the Harz district. First encountered in his twenties and newly married, he lived at Eisleben in Saxony. He probably worked as a laborer, and seems to have been very poor. Two years later he moved to Mansfeld, not far away but deeper into the mountains. Here the rewards in mining were greater than in Eisleben; he was still poor, but in 1491 he was evidently rising in the world, for then he was elected to Mansfeld city council. By 1502 he had leased some mines to run on his own account. It was a five-year lease, but it was renewed in 1507, by which date he had become part owner of six shafts and two furnaces.

This particular miner is well known because his second son was Martin Luther, born at Eisleben in 1483. Martin attended the university at Erfurt, and after completing his course in 1505, he became a monk in the Augustinian house there. In this capacity he later went on to Wittenberg University in Saxony as a teacher; it was there that he developed his criticisms of the Church, and there in 1517 that he launched his attack on it.

The connection of Martin Luther with the mining industry was more than just the accident of his father's occupation; living in Saxony, he was not far from some of the most important mines, and the mining communities identified themselves closely with his teachings. Thus Agricola's personal acquaintances included two Lutheran pastors who wrote books on mining, partly with the aim of popularizing Agricola's work. One was Johann Mathesius (1504–65), who had charge of a church at Joachimsthal. His comments on mining and metals appeared in a book of sermons called *Sarepta,* after a town mentioned in the Bible.[11] Mathesius compared this place to Joachimsthal because its Hebrew name, Zarephath, meant "place of refining."

Another book of sermons by Mathesius, published in 1566, is of importance to historians because it contains the first extended biography of Luther. As a child, Luther had a good home and was well provided for, Mathesius said, because "God blessed the mining industry."

Other pastors preached in this way, and there was evidently a tradition of sermons illustrated by mining and metallurgical information that continued into the seventeenth century. There were even hymns written on this theme; Agricola had noted that the miners "lighten their long and arduous hours by singing, which is neither wholly untrained nor unpleasant."[12] Luther himself wrote hymns and took immense pains over the musical settings of his German Mass. Indeed, he originated a musical tradition whose "greatest exponent" was the incomparable J. S. Bach. And he said that whoever "despises music, as do all the fanatics, does not please me. For music is a gift and largesse of God. . . . Music drives away the devil and makes people happy."[13]

Neither Luther nor Agricola said anything about the words of the miners' hymns and songs, but a hymn published in the seventeenth century spoke of a great mine owner, "born of David's line, a Lord of the whole land," who chose to go down and work in the mines himself. He broke through the rock in a shaft called:

. . . little Bethlehem,
looked for a gallery,
and went to Jerusalem,
. . . the gallery had
treacherous stone in it,
little ore could be seen,
that was of rich content . . .[14]

This kind of identification of the mining communities with Lutheranism was reinforced on a theological level by Luther's doctrine of "the priesthood of all believers," first put forward in 1520. This was an attack on the Church's notion that priests had a special "vocation" or calling from God, which people in other occupations did not share. According to Luther, every calling was of God, and a farmer or miner who tried to follow Christ in doing "his work faithfully . . . is equal to the priest in God's sight." Each man, peasant or cleric, who helps his fellow man in Christian charity is equally regarded by God's standards. So however naïve the hymn just quoted, which pictured Christ working in the mines, it did express a thought that was very characteristic of the Lutherans.

Insofar as Luther's teaching implied that a man's ordinary daily work could be pleasing to God, it had some similarity with the "work ethic" of the Cistercian monks, discussed in chapter 1. As Luther had himself been a monk, he must have known something of the belief that manual work should play a part in monastic life. This was something St. Benedict had taught in the sixth century, which had influenced most European monastic orders to some degree. A particular stress had been laid on this principle by the leaders of the Cistercian order, especially St. Bernard of Clairvaux. Partly as a result of the high value they accorded to manual work, Cistercian monks had built mills, smelted iron, and dug mines. Whether Luther knew very much about all this is doubtful, but he certainly "venerated Bernard,"[15] and his theology was influenced by Bernard's writings on St. Paul. And although Luther despised the life led by most monks in his own day, he thought that if it was pursued in the spirit of St. Bernard, monastic life could be of great value.

The implications of this were that mining should be more highly valued as work approved by God. Agricola also wanted mining to be more highly regarded, but mainly in an academic way. He was a scholar with a Renaissance outlook, and it was for men of learning that he wrote, not miners. Thus the preacher quoted earlier, Mathesius, found it necessary to interpret Agricola to his miner friends. At one point he commented that he was running "through the works of Agricola . . . looking for certain things that may be of interest to you."[16] Here was a cultural gap that Mathesius felt the need to bridge. The degree of his success is uncertain, for by the end of the sixteenth century the gap was being filled by a very different kind of technical literature, the author of which was principally Paracelsus. This was the pen name of a German-speaking Swiss whose real name was Bombastus von Hohenheim. In the course of a turbulent life, he wandered through many of the German mining districts and other parts of eastern Europe. He is said to have served an apprenticeship in a mine, and he worked for a time as an army surgeon. His reputation as a medical practitioner was established when he performed an apparently miraculous cure for Johannes Froben, the famous Basle publisher associated with Agricola and Erasmus. On the strength of this he was allowed to lecture at Basle University for a year, until his controversial manner led to his expulsion.

Paracelsus wrote extensively on a wide variety of subjects, but little of his work was published until after 1570. Then his books attracted much attention, and by 1600 they were being widely read by those who sympathized with his frequent attacks on the accepted authorities in medicine and alchemy. His reputation as a reformer of these subjects was such that he came to be regarded as the "Luther of medicine." In one passage he commented that the folly of books influenced by Aristotle had incited him "to write a special book concerning Alchemy, basing it not on men, but on Nature herself, and upon those virtues and powers which God, with His own finger, has impressed upon metals."[17]

The comment about "virtues and powers" reflects the influence of neo-Platonism, which had earlier enjoyed a renewed popularity in Italy. In other writings, Paracelsus claimed that his theories proceeded from "the light of Nature." This too sounds neo-Platonic, but its general tone also coincided with the view of some of the more extreme Protestants who claimed that ordinary people, informed by an "inner light," had as much access to God's truths as had a priest or a man of learning. This view arose partly from a distortion of Luther's ideas, for where Luther had merely said that all vocations could be equally worthy, it was now being claimed that all men had equal access to knowledge, through inspiration, the inner light, and, in technical matters, through the intuition and experience acquired by craftsmen.

Such views were held by the Anabaptists and by some of the leaders of the Peasants' Revolt of 1524–25, who thought that ordinary people, with no theological training, could take the reform of church and state into their own hands. Luther was appalled by this; he valued scholarship and was in many ways conservative in his approach to reform. But his sharp condemnation of the Peasants' Revolt led to much bitterness among people who had thought of him as their leader.

It was the persistence of these more extreme kinds of Protestant opinion that provided a receptive audience for the books of Paracelsus as they came from the presses at the end of the sixteenth century. There were plenty of people ready to be told that conventional learning had "to have its ineptitude propped up and fortified by papal and imperial privileges,"[18] and that knowledge of chemistry was "graven . . . in the metals," and was to be acquired not from books, but by craftsmen as they worked and through inspiration stimulated by experiment. Technical knowledge was the product of a spiritual process and of manual work, and did not arise from intellectual or scholarly researches.

Paracelsus wrote about the diseases of miners, but not about the mines themselves. His main interests were in medicine and chemistry. He had theories about salt, sulfur, and mercury, the three chemical "principles," and he had opinions as to how the influence or virtue of the planets could act through metals. His ideas were thoroughly mixed up with alchemy, so for the modern reader it may seem disappointing that they should have had greater influence than the careful scholarship and the balanced empiricism of Agricola. But in contrast to Agricola's learned Latin style, Paracelsus wrote in the German vernacular, in an idiom that touched the imagination and idealism of ordinary craftsmen. He identified himself with the "lower social groups" and with some of their Protestant attitudes, although he himself was always nominally a Catholic.

Agricola had more in common with the intellectuals of Lutheranism than with the ordinary miners, and he lived as a highly respected citizen of the predominantly Lutheran mining town of Chemnitz. Nonetheless, he remained staunchly Catholic, though his churchmanship was of a very different stamp to that of Paracelsus. In the 1520s he had written some tracts on religious topics. He was at that time a critic of the papacy, and like his friend Erasmus he may have sympathized with Luther's early demands for reform. But with the widening breach between Luther and the Church, and the emergence in 1524 of the Anabaptists, many of those who thought like Erasmus turned away. They disliked Luther's vehemence and apparent sacrifices of reason to faith, and wanted instead a moderate and tolerant Church that stressed morals more than mysticism.

But despite his disagreements with them, Agricola was greatly appreciated by the Lutheran leaders. When he died, one of them wrote, "I have always admired the genius of this man, so distinguished in our sciences and in the whole realm of Philosophy—yet I wonder at his religious views."[19] Unfortunately, this admiration did not give Agricola's books an influence as great as the writings of Paracelsus.

Commerce and Warfare in Germany and England

To understand the growing interest in the writings of Paracelsus and the earlier identification of German miners with Luther, we need to recall that the sixteenth century had been a period of expanding trade and "commercial revolution" in Europe. One reason for this was the shipments of bullion coming from the Americas and trade on the new sea routes to the East Indies. Another reason was the removal of constraints on trade formerly exercised by craft guilds, the church, and the monasteries, releasing brakes that earlier had held back enterprise. Merchants and the owners of estates were increasingly able to use their assets for their own individual pecuniary advantage without reference to wider social obligations. For example, landowners could more easily enclose large tracts on their estates regardless of the effect on peasants who had formerly grazed their animals there but were now excluded. Merchants could more often disregard guilds that attempted to protect the interests of craftworkers. Moreover, in a period when prices were rising, miners, artisans, and peasants often found themselves worse off. Several processes were at work, therefore, in various parts of Europe (including Britain), which caused poverty to increase even while prosperity was generally growing.

England began the sixteenth century as one of the more backward countries of Europe, but experienced a more rapid expansion than most others, particularly when the dissolution of the monasteries in 1539 freed land and financial resources for new uses. In the next hundred years, England became one of the leading mine-working nations of Europe, with the copper and tin mines of Cornwall growing in importance, and with coal mining increasing to an extent very unusual elsewhere in the world at the time.[20]. The manufacture of woollen cloth also increased dramatically, especially in Pennine hill villages where there were no guilds. England had once been a major exporter of raw wool, but now it sold cloth on the international market. Technological innovations associated with these developments included the use of coal as fuel in brickmaking and a machine for knitting stockings, invented by William Lee in 1589, as well as the improvements in mining mentioned earlier, including simple wooden railroads on which horses pulled cartloads of coal. Both the latter inventions were first used near Nottingham, an important industrial area where members of the local mine-owning Willoughby family were also interested in mathematics and science.[21] Lee's hand-operated knitting machine did not enjoy immediate success, but after 1650 increasing numbers were used in the Nottingham stocking manufacturing industry.

Some historians see these developments in the sixteenth century as closely connected with the Protestant reformation, not only because it had released resources formerly held by the monasteries and the Church, but also because some brands of Protestantism provided positive ideals for economic enterprise. This does not apply to Luther and the somewhat extreme Anabaptist sects, who were basically hostile to the bankers and merchants of the new order. The Anabaptists, in particular, became identified with artisans and craftworkers disadvantaged by current economic developments. However, John Calvin (1509–64), in Geneva, was one religious reformer who was less negative in his attitude to commerce, and rather than condemning it outright suggested moral disciplines through which its excesses could be controlled. In its early form, as practiced in Geneva (and rather later in Massachusetts), Calvinists exercised this discipline *collectively* through the Church community. Only later was this approach replaced by an ideal of personal responsibility and self-discipline more compatible with business life.

This latter version of Protestantism had a strong appeal for Dutch businessmen as well as English puritans, and possibly provided a stimulus for industry and technology as well as commerce. The case for this interpretation has been strongly argued in the well-known writings of Weber and Tawney,[22] while among historians of technology, Friedrich Klemm has been particularly emphatic about Calvinist influence.[23] However, the kind of religious idealism that was more prominent before 1650, and with which this chapter is concerned, is not the "Protestant ethic" with its justification for economic enterprise; it is the rather different ethic of thinkers and reformers who, like Paracelsus, sympathized with artisans and peasants, or were concerned about the increasing numbers of the poor. We will encounter Protestant entrepreneurs and industrialists later. For the moment we are concerned with a religious view that was more emphatic about charity than commerce, arguing that the proper role of knowledge, discovery, and invention should be the benefit of humankind as a whole through improvements in medicine, agriculture, and "husbandry."

This social ideal for technology gained ground when the economic expansion of the sixteenth and early seventeenth centuries was interrupted by a period of warfare and depression that has been termed "the general crisis of the seventeenth century."[24] In the Low Countries there was a long war against Spanish domination beginning before 1600. In central Europe there was the Thirty Years' War (1618–48), and in England the Civil War (1642–49). To the devastation caused by these conflicts was added a series of plagues that killed many people in Italy and Germany from the 1620s onward. According to one estimate, the German states as a whole lost 40 percent of their population in the three decades leading to 1650.[25]

The way in which these wars encouraged thought about the redirection of technology toward human welfare can be illustrated by the career of Rudolph Glauber (1604–70), a German who traveled widely in Europe during the war period. He wrote about chemistry and medicine, discovered the medicinal properties of sodium sulphate (since known as Glauber's salt), and became interested in the works of Paracelsus. The poverty of war-torn Germany impelled him to study the practical arts in relation to the reconstruction of trade and the better use of the metals from German mines. He had a special interest in agriculture and advocated the use of saltpeter as a fertilizer, commenting that "fruits will then ripen much earlier and have a much pleasanter taste than if stinking cow-manure had been used." Elsewhere he wrote that although Germany had many mines, it lacked "experienced persons who know how to use them properly, and then to find timber and sufficient of all other necessities." This indicates how far the German metal industries had declined during the war, as does Glauber's comment that unworked metal produced in Germany was being exported to Holland, France, and Venice. There it was used in the manufacture of various goods and materials, some of which were being bought back by Germans at greatly enhanced prices. Germany had all the raw materials it needed for any sort of manufacturing. "Why are we so bad at it?" Glauber exclaimed.[26]

Glauber's writings on "Germany's welfare" are filled with a concern to help his fatherland. He discussed alchemy, quoting Paracelsus. A large number of his ideas were highly speculative, but his aims were not in doubt. Many other people at this time studied technical subjects with similar objectives. Among them there grew a sense of the importance of useful knowledge, and, accompanying this, an appreciation of the need for new kinds of education, in which practical subjects were given more attention.

One of the leading exponents of the ideal of practical education was J. A. Komenský (1592–1671), better known by his Latin name, Comenius. He was a minister and bishop of the "Unitas Fratrum," a small sect with ideas somewhat like the Lutherans', but with a longer history. Known in more modern times as the Moravian Church, this body had been formed in Bohemia during the 1450s and had survived in that part of Europe until 1621, when the advance of Catholic armies during the Thirty Years' War almost wiped it out. Before 1621, Comenius worked in a school at Fulnec in Moravia, but with invasion impending there, he fled to Leszno in Poland where he again took up his work as a teacher.

Comenius believed that knowledge, mastered and shared, could change the world. Hence the aim of much of his work was to improve education. He also believed in the unity of all knowledge and wrote books of an encyclopedic kind to illustrate this. Knowledge was to be gained both from the study of nature and from the scriptures. In a treatise on "Natural philosophy reformed by divine light" (the significant title of its English translation), he argued that science should be based on experience, reason, and scripture; in particular, he thought, the creation of the world as it was described in Genesis gave support to the ideas upon which Paracelsus had founded his chemistry.

The books Comenius wrote on subjects connected with education and the extension of useful knowledge attracted wide attention. They were welcomed enthusiastically by English Puritans and were published in English translation from 1631 onward. England and Sweden were the countries most admired by the Comenians—Sweden was thought well of because its military power was the most effective force on the Protestant side during the Thirty Years' War. In addition, Swedish mines and industries were developing in an exemplary way under the management of Louis de Geer, a Calvinist banker who himself greatly admired Comenius and sought to put his educational ideas into practice. England, a country where the Protestant cause seemed secure and where Protestant refugees from eastern Europe were settling, was also admired because it had been the home of Francis Bacon (1561–1626), the most eloquent and influential of all the advocates of "philanthropic" science and socially useful technology.

Social Ideals for the Use of Knowledge

Although Bacon died in 1626, his ideas were studied with renewed interest in the 1640s and 1650s. Many people at this time habitually referred, like an earlier generation (chapter 2), to the last book in the Bible and believed that the warfare of their era heralded the end of the world as they knew it. But their interpretation was optimistic in expecting that this crisis would be followed by the dawning of a new era. The invention of printing, the discovery of the Americas, and the continuing growth of practical knowledge all seemed so remarkable that people felt that a decisive change in human affairs would soon occur. It would be a dramatic change for the better: the attainment of a utopian millennium, or the recovery of a lost golden age. Thus it is significant that the series of books on the advancement of knowledge that Bacon planned and partly wrote was to be called "The Great Instauration," indicating the dawn or instauration of a new phase in the history of humankind.[27] The main volumes in this series were to be a new edition of his early book *The Advancement of Learning* (1605), followed by books on two key aspects of scientific method—a new logic and a natural history.[28]

At the very start of this series, Bacon attacked the misuse of knowledge and stressed a social ideal for the application of science in what we would now call technology. Knowledge and skill should develop only to bless the life of humankind, he argued. If knowledge were "severed from charity, and not referred to the good of . . . mankind," it would have an "unworthy glory."[29]

Bacon wrote two other books of an even more visionary nature that were published together in one binding in 1626, just after his death.[30] One was partly based on an earlier work dealing with the nature-defying art of "natural magic" written by the Italian writer G. B. della Porta (chapter 3; in figure 21 the actions of the devices illustrated would be regarded as "magic"). Bacon aimed to disentangle the practical knowledge included in this subject from the "fantastical learning" that surrounded its discussion. But ideas about magic were significant because they had raised people's expectations of technology, and Bacon's optimistic view of technological potential was fed by these ideas also.

Published with this book was the famous utopian work, *The New Atlantis,* which was Bacon's vision of what life would be like when the "Great Instauration" was completed. It gave details of an institution named "Solomon's House" dedicated to scientific and technical researches, and indicated quite clearly that knowledge in these subjects was only gained through patient and careful inquiry and experiment, not simply by inspiration. Most interestingly, it indicated the kinds of subject to which research should be devoted. Gardening, agriculture, and baking were topics where innovation would obviously be of benefit to humanity. But there would also be research on flying machines, submarines, and methods of refrigeration, subjects related to "natural magic" on which several of Bacon's contemporaries speculated.

The link between Englishmen who admired Bacon and eastern Europeans like Comenius was provided by Samuel Hartlib (1592–1662) and John Dury (1596–1680), both of whom had lived at Elbing in Polish Prussia until forced to leave by the advance of the Catholic armies. Hartlib came to England in 1628 and for two years ran a school on Baconian principles in Chichester. It was not a success and he soon moved to London, where he applied himself to helping refugees from Germany and eastern Europe. One of these was Johann Christoph de Berg, an inventor who had worked on mine-drainage problems in Moravia. Dury meanwhile traveled widely in Europe, campaigning for unity of all Protestants and carrying messages for Hartlib. Dury's function was partly to oil the wheels of Hartlib's vast European correspondence. The latter wrote so many letters and his contacts were so wide that he became known as "the great intelligencer of Europe." He put himself forward as a universal secretary who would link good men in many countries through an "office of address." The aim would be to improve agriculture, teach languages, forward inventions, compile statistics, educate the poor, and generally work for the good of all. Much could be achieved in this way if a united, tolerant Protestant society could be established.

The campaign for educational reform supported by Hartlib and many of his Puritan friends came to a head when Parliament debated the proposals in 1653. Among those outside Parliament who were provoked into discussing the question was John Webster, a former chaplain in Cromwell's armies. Webster's ideas were less realistic than Hartlib's but well illustrate the educational ideals of the more radical Puritans. In a series of sermons preached in 1653, Webster criticized the traditional teaching of the universities. "The Apostles and Disciples," he said, "bad us beward of Philosophy, which is after the rudiments of the World, and not after Christ." Philosophy at the universities is carnal and frivolous: "they have atomised the unity and simplicity of . . . truth . . . accumulating a farranginous heap of divisions and sub-divisions, distinctions, limitations, axioms, positions and rules, (which) do chanel & bottle up the water of life."[31] In the study of "natural philosophy," Webster argued, Aristotle's works should be forgotten, except for his "History of Animals," but new materials should be introduced from the works of Descartes, Paracelsus, van Helmont, and Gassendi.

In all this, he emphasized, the knowledge to be encouraged was that which could be used for "the general good and benefit of mankind, especially for the conservation and restauration of the health of man."[32] At the beginning of another of his works, *The Saints' Guide,* Webster actually said that learning by itself was a sin, "but if it be sanctified . . . the providential wisdom of God doth . . . make use of it for the good of his People." So, for example, academic mathematics was to be condemned, but the universities should teach accountancy, the mathematics of navigation, and the geometry that a carpenter might use.

But the stress on useful and practical studies was made most fully in relation to the chemical arts. Students should "learn to inure their hands to labour, and put their fingers to the furnaces, that the mysteries discovered by *Pyrotechny,* and the wonders brought to light by *Chymistry,* may be rendered familiar unto them; that so they may not grow proud with the brood of their own brains, but truly be taught by manual operation, and ocular experiment, that so they may not be sayers but doers."[33] Webster went on to talk about "nature's hidden secrets," and he was clearly reflecting the Paracelsian view that knowledge was to be acquired through the light of nature, which was made accessible in the course of working at a trade or making a practical experiment.

Although he had studied medicine and eventually settled down to practice as a physician, Webster also showed what he had learned by putting his hand "to the furnaces" in a very competent and informative book on metals. Entitled *Metallogra-*

phia and published in 1671, it was subtitled in Baconian style "an history of metals" while at the same time drawing inspiration from Paracelsus. The book makes it clear that Webster had visited lead mines in the part of northern England where he lived, had collected samples of ore, and had worked with a local assayer who determined the silver content of the lead.[34]

While Hartlib would have approved this book, he was unsympathetic to the widespread Puritan antagonism to secular learning that people like Webster expressed, but he did share the ideal of providing a more practical form of education. To some extent this ideal was realized with the foundation of a college at Durham in 1657, based on plans made by some of Hartlib's associates and made effective by the personal support of Cromwell. The college was to teach languages, chemistry, agriculture, medicine, and practical arts. Language teaching was to be based on the methods of Comenius and there was a workshop for teaching the practical subjects. The college was forced to close after the death of Cromwell in 1659, that is, after only about eighteen months of operation. But its ideals survived to some degree in the Dissenters' Academies of the late seventeenth and eighteenth centuries, and the general emphasis on practical and useful sciences was shared to a limited extent by the Royal Society, chartered in 1662.

Since much of the impetus for these ideals concerning practical education had come from eastern Europe, it is not surprising to find a similar educational movement developing there. The ideas of Comenius were taken up with enthusiasm by the Pietists, a group within the Lutheran churches that, from the 1670s onward, emphasized a "religion of the heart" and the inward spiritual life. One of the Pietist leaders, August Francke (1663–1727), was closely involved with the foundation of the university of Halle in 1693–94 and became a professor there. Halle was the first German university to introduce studies in the sciences on a really large scale, and Francke's former students founded secondary schools that taught practical subjects at Halle (1708), Berlin (1747), and other places. The teaching of science at Halle also influenced its adoption soon after at other universities, of which Göttingen is one example.

But as with other seventeenth-century movements that emphasized practical education, abstract mathematical studies were often suspect at Halle; the rationalism of the "scientific revolution" was sometimes regarded as irreligious, and there was a tendency to adopt a somewhat spiritual view of chemistry and the practical arts. The first professor of chemistry at Halle, appointed in 1693, was G. E. Stahl (1660–1734). He had believed in alchemy while in his twenties, but although he became more skeptical later, his phlogiston theory of combustion is regarded by many historians as one of the last phases of "prescientific" chemistry. However, we should be wary of such judgments, not only because the phlogiston theory had wide influence on ideas about the chemistry of combustion. A further point is that the "spiritual view" of chemistry (and metallurgy), with its strange-sounding statements about alchemical "secrets of nature" (or knowledge discovered by the "light of nature") is a view concerned with real experience in the lives of craftworkers.

We can understand this with the help of C. S. Smith, the historian of metallurgy, who draws a contrast between those who attempted to explain the natural world "in terms of structures . . . rather than in terms of quality." In the former camp were men like Galileo and Descartes who analyzed problems in terms of mechanics, who thought of matter in terms of structures built of atoms, and who dismissed color and texture as "secondary qualities." By contrast, the metalworker with his furnace, his crucibles, and his anvils was conscious mainly of quality as indicated by the color and brightness of heated metal and its consistency and ductibility. He fashioned the metal according to his "feeling" for quality, guided by what were primarily aesthetic judgments as to whether the glowing material in the furnace looked right for the next stage in the process. It was this feeling and aesthetic sensibility that was being referred to when references were made to a "spiritual" aspect of craftwork, or to the "light of nature."

Technology is more than just applied science because it depends on awareness gained during practical work, not only on abstract knowledge. We admit this in the twentieth century when we talk about the importance of "hands-on" experience, a phrase that echoes John Webster's wish that students should "put their fingers to the furnaces." But it was abstract knowledge of machines and mathematics that came to be associated most strongly with discovery in science and the refinement of engineering theory. One result was that "extensive but purely qualitative knowledge . . . of matter that existed in the seventeenth century was partly forgotten."[35]

Technology and Human Destiny

Reviewing what has been said in this and previous chapters, it is tempting to classify attitudes to technology during the sixteenth and seventeenth centuries in terms of three distinct movements. First, there was the interest in mathematics and engineering discussed in the previous chapter and exemplified strongly in Italy and France by men such as Galileo and Mariotte. In the Low Countries, we may now add, this tradition also developed strongly, stimulated by the work of instrument makers as well as drainage engineers. Two leading writers on mathematics and mechanics, comparable in their relationship with engineering to Galileo, were Simon Stevin and Christian Huygens. A second major movement in the period was the enterprise in mining, and also textiles and foreign trade, associated with German bankers such as the Hochstetters, or with industrialists from the Low Countries such as Louis de Geer. Third, there were the artisans, craftworkers, miners, and peasants who claimed direct access to knowledge through practical experience and who seem to have found their attitudes expressed to a large extent in the mysticism of Paracelsus.

Needless to say, these different trends are not always sharply distinct, and contrasts between them are in any case blurred by the fact that the people who wrote about technological matters were rarely themselves either artisans, entrepreneurs, or mine captains. Indeed, several of the authors quoted were medical men (Agricola, Paracelsus, Webster), Catholic clergy (Kircher, Schott, and Mariotte—see chapter 4), Protestant ministers (Comenius, Mathesius), or members of other professions (Bacon the lawyer, and Galileo, a university teacher). Some of these men had to advise on or supervise engineering projects (Galileo, Stevin, Mariotte), and a few of them, most notably Galileo, showed some awareness of entrepreneurial attitudes and skills such as cost accounting.

Connections between movements in technology and the various Protestant churches and sects are worth noticing. The strongest and most significant connections are those involving artisans and craftworkers. In Germany, as we have seen, many miners became Lutherans. In Holland, artisans settling in the major towns were often Calvinists, Anabaptists, or Mennonites. In England about 1600, Puritanism was very strong in cloth-manufacturing districts and among the "small masters" who owned weaving businesses.

In none of these countries, however, did Calvinist entrepreneurs operating on any scale emerge until the seventeenth century.[36] Prior to that, the most prominent bankers and merchants in London, in Bruges, and in other parts of the Low Countries were Italians. We have also noted the importance of German bankers in many parts of Europe. Most were at least nominally Catholic. Moreover, it is striking how much of the expertise needed in commerce as well as in technology had originated in Italy, largely during the fifteenth century. Previous chapters have demonstrated this for ideas about machines. Now we need to note that business methods used by German, Dutch, and English merchants in the seventeenth century had been learned from Italians in the previous generation, included techniques for banking and insurance, credit and currency exchange, and the approach to cost accounting that we have seen reflected in Galileo's work (chapter 3) and in Hochstetter's assays of metal ores. In a quite different sphere, even Paracelsus drew many of his ideas from Italian neo-Platonist writers.

The fact that these ideas were now being developed more actively in northern Europe had much to do with developments in overseas commerce. If trade was stagnant in Italy but expanding in the North, this was mainly because new trade routes had bypassed the Mediterranean and brought more and more ships first to Antwerp, then in the seventeenth century to Amsterdam and London.

It would seem then that artisans and small tradesmen identified themselves with Protestantism at an early stage, but that the Calvinist entrepreneur stressed by some historians emerged rather later, mainly after 1650. It also seems that an explicit "Protestant ethic" favorable to business and commerce only emerged in theological writings in this later period. Calvin was less hostile to bankers and merchants than Luther, but only in the sense of making "a qualified concession to practical exigencies." Thus R. H. Tawney stresses that the so-called Protestant ethic was not the work of Calvin himself but of later Calvinists who discovered in their religion "a frank idealization of the life of the trader, as the service of God and the training-ground of the soul." Tawney notes that his eminent predecessor, Max Weber, failed to recognize this development in Calvinist thinking and hence failed to see that "the individualism ascribed not

Table 1
Some idealistic trends in technology before 1700, discussed in chapters 1–5

(1)	Technical ideals — concerned with explorations *within* technology
a)	*Intellectual idealism* emphasizing: — *mathematics,* measurement, and numbers (chapters 2 and 3) — *machines*: Galileo (chapters 3 and 4) — *rationality* in analysis and organization (chapter 4)
b)	*Aesthetic ideals and symbol creation* — in *building* and architecture (chapters 1, 2, and 3) — in *metalworking* (chapter 5) and clockmaking (chapter 2)
(2)	Social and other ideals — concerned with *uses* of technology
a)	*Social ideals* — *charity* and service to humanity: Bacon (chapter 5) — the *value of work* (chapters 1 and 5)
b)	*Economic goals* — medieval commerce (chapter 1) and the idealized commercial life of the Protestant ethic (chapter 5)
c)	*Creation of symbols* representing: — the status of cloth towns (chapter 1) — the political power of the Church (chapter 1)

unjustly to the Puritan movement in its later phases" would have "horrified" the early Calvinists.[37]

These developing ideas in theology are relevant not only to the business ethic of the *late* seventeenth century but also to ideas about human destiny and the role of technology and industry in that destiny. For the later Puritan, hard work in industry or trade was a private responsibility related to his individual salvation. For Bacon, Comenius, and other *earlier* seventeenth-century Protestants, industry, trade, and the advancement of knowledge were to be pursued for their social benefits, not for individual "profit, or fame, or power," but inspired by compassion for suffering humanity. And the approaching millennium would be a great new dawn for all humankind, not just as the summed salvation of many individuals. For Bacon, the development of technology was a social enterprise, carried out by many people working cooperatively, notably in the "Solomon's House" research center in *The New Atlantis.* And he also stressed the benefits of wealth in social terms, arguing that the greatness of the nation would be most advanced if "treasure" was "rather in many hands than few."

Bacon expressed his ideal for what we now call technology in language drawing on the Bible in almost every sentence and oriented toward Biblical teaching on charity, as well as on salvation. But his comprehensive and moving vision of the potential for applied knowledge has been a source of inspiration for many in later generations. He was quoted almost like scripture by later Puritans, but his ideas could also be interpreted in a secular way. And in this sense, too, he defined ideals and goals, indicated methods for improving "trades," and suggested institutions through which improvement and development might be promoted. Bacon is also sometimes criticized, notably by modern environmentalists, for his use of one other Biblical idea, that of man's proper role in "dominion over nature." However, the overwhelming emphasis of his work was compassionate, and he was uneasy with doctrines that claimed enlarged scope for the free play of economic self-interest. So in the divergence between his hopeful ideal of universal benefit and the kind of Puritanism concerned with the sanctification of business life, we can see once again the contrast between social ideals and economic goals in the development of technology (table 1).

6

The State and Technical Progress: 1660–1770

French Policies under Colbert

Seven or eight years ago I showed to Monsieur Colbert an engine which I had built . . . It worked as follows: a tiny quantity of gunpowder, about a thimbleful, was able to raise some 1,600 lb. . . . with a moderate and steady power. Four or five servants, whom Monsieur Colbert ordered to pull the rope attached to the engine, were quite easily lifted into the air.[1]

So wrote Christian Huygens in 1686, describing experiments that have already been discussed in chapter 4. The incident illustrates the close interest in science shown by one of the most important ministers in the French government, Jean Baptiste Colbert.

Colbert served Louis XIV as *Contrôleur général des finances* from 1661 until his death in 1683, and in this office he had a general responsibility for the economic welfare of France. A problem that concerned him particularly was the inadequacy of French shipping in relation to the huge expansion of overseas trade that some nations were enjoying, and he seized every opportunity for making improvements not only in the number and size of ships but also in the training of their officers, in the preparation of charts, and in the art of navigation.

Colbert probably first became interested in Huygens's work because of a possibility that pendulum clocks designed by Huygens might be adapted for use as chronometers on ships. These would have been of the greatest value in navigation, because they would have provided an easy way of determining longitude. But despite many ingenious designs and much experiment, Huygens was not, in the end, successful in his attempts to make a chronometer for use at sea.

For Colbert, though, navigation was just one of many practical subjects where mathematics and science seemed potentially useful to the state. For this reason he encouraged mathematical and experimental studies in France as a matter of government policy. Colbert knew that in London the Royal Society had been chartered in 1662, and that one of its aims was to try to improve trades and the practical arts. In 1666 he founded a somewhat similar body in Paris, the *Académie Royale des Sciences.* Huygens was well known in France, although his home was in Holland, and Colbert encouraged him to take office in this body. Huygens accordingly did so and "was indisputably the one who chiefly guided the affairs of the Académie".[2] From 1666 until 1681 he lived almost continuously in Paris, returning home to The Hague only for short visits.

In most respects this was in no way remarkable, for Huygens was the son of a diplomat, and the family had been used to living away from home. Descartes had been a regular visitor to his father's house; there were many family friends in Paris. But as the policies of Louis XIV and Colbert developed, they became increasingly hostile to Holland, and Huygens was criticized to some degree for appearing to help an enemy.

The greatest cause of tension between France and Holland was the preeminence of Dutch shipping. Much of France's trade was being carried in Dutch vessels, and it was felt that the prosperity of the country was being endangered by the concentration of so much traffic in Dutch hands. In 1669 Colbert himself drew up a memoir to underline the point. He estimated that European owners altogether possessed about 20,000 ocean-going ships—that is, ships of 60 tons burden or more. Of these, 15,000–16,000 were owned by the Dutch, 3,000–4,000 by the English, and only about 500 belonged to French owners. Colbert believed that there was little scope for increasing the total volume of seaborne trade, so if France was to increase the size of her merchant fleets, as was necessary for her prosperity, the size of the Dutch fleet had to be reduced. Commerce was a form of warfare, and the Dutch were the principal enemy. Whatever Huygens could teach the French about navigation would be an asset here; with regard to shipbuilding, Colbert persuaded a number of Dutch carpenters to work in French shipyards and teach the local craftsmen by their example. He commissioned a scientific analysis of hulls of different shapes with a view to finding the best combination of factors relating to speed, maneuverability, and the positions of gun ports. In addition, he founded schools of naval construction at Brest and Rochefort, and at the same ports established academies for training naval officers (a three-year course) and artillery officers. Rochefort and Dieppe were also given schools of hydrography. Thus technical education played a major part in naval policy.

Another aim of Colbert's was to promote the formation of trading companies. In 1670 he was able to claim that by competing with the Dutch for trade between France and the West Indies, these companies were depriving the Dutch of eight or ten million livres a year (the equivalent of something like £2 million a year in modern currency).

Huygens was far more interested in mathematics than in national rivalries, and Paris provided him with the best available environment in which to study. But he continued to live in Paris even when Colbert's policies led to war between France and Holland in 1672. And although Huygens returned to Holland in 1681 as a result of ill health, he only ceased to be welcome in Paris after the death of Colbert in 1683, and after official toleration of Protestants and Jews was ended in 1685 by the Revocation of the Edict of Nantes.

Colbert's policies for the economic prosperity of France had far-reaching implications for the whole internal economy of the country. There was an effort to promote new industries and to reorganize existing ones. An edict of 1674 forced all trades to be administered by guilds, on which taxes were levied and which had to enforce regulations about bookkeeping, company formation, and apprenticeships. Many businessmen hated Colbert for his interference, but others recognized his fair dealing and genuine support for their work and called him "the protector of merchants."[3]

Among Colbert's most successful and lasting work for the French economy was the improvement of inland transport. He initiated several small road-building projects, but no funds were available for roads after the start of the Dutch war in 1672, and it was with river and canal transport that most was done. France was well endowed with naturally navigable rivers, which had been extensively used from a very early date, but by building weirs and locks, by digging new channels for some lengths of river, and by dredging others, it was now made possible for sizeable boats to navigate the upper reaches of streams previously impassable. Work of this kind was undertaken on the

Rivers Seine, Aube, Loire, Rhône, Marne, Somme, and several others, and the Orleans Canal was made between 1679 and 1692 to provide a link between the Loire and the Seine.

But all these schemes were on a very small scale compared with the project to build a "canal of the two seas," which would be some 175 miles long and would link a river that flowed into the Atlantic with one flowing into the Mediterranean. By this means, traffic for Mediterranean ports would be saved the long journey via Gibraltar, and the Dutch would lose even more of their seaborne trade. The idea of such a canal had long been discussed, but it became possible only when Pierre-Paul Riquet demonstrated how water could be supplied to the canal's summit level, which was about 600 feet (180 meters) above sea level. Riquet performed experiments to test the feasibility of his ideas and eventually designed a water-supply system that included a large impounding reservoir and water channels extending for some miles.

Riquet wrote to Colbert about his scheme in 1662, suggesting that the Crown should buy the land needed for the canal and finance its building; when it was finished, he would operate the waterway and maintain it with money raised in tolls. Colbert greatly approved of the scheme, not only for its economic value but as a monument to French achievement under Louis XIV. With this in mind he wrote to Riquet stressing that the works should be built to the highest standards.[4] Riquet certainly did build well and proved himself also an expert in organization, for at times he was using a labor force of nearly 10,000 men. About a third of the cost was borne by the Crown, about a fifth by Riquet himself, and the rest by the provincial authorities. The canal was completed in 1681 and was soon carrying a considerable traffic. But the cost of trans-shipping cargoes from seagoing ships to canal boats prevented it from becoming a real alternative to the sea route via Gibraltar.

Developments in England

It was another century before England had a waterway that could compare with the Languedoc Canal, as the "canal of the two seas" was more properly called, but contemporary with the improvement of navigable rivers under Colbert there were comparable, if smaller, river schemes being promoted in England. Shortly after the Restoration, Parliament passed Acts to allow a number of small rivers to be made navigable—the Wiltshire Avon passing through Bath (1665) and others. Then in 1699 there was the important scheme based on the River Aire to link the Yorkshire towns of Leeds and Wakefield with the sea at Hull. The Trent was made more effectively navigable up to Burton, and a link with Derby followed. In 1720 an Act was passed to make the Rivers Mersey and Irwell navigable up to Manchester.

There were several similar projects at this time, and altogether a great deal was done during the two or three decades before 1730. The period was one of vigorous material progress in other ways as well. The output of tin from the mines of Cornwall was rising fast, and the smelting of Cornish copper, previously a very small operation, was increasing even faster. This in turn helped to stimulate the improvement of pumps for use in mines. From 1712 Newcomen's engines were available for this purpose, but these machines found their most extensive applications in the coalfields where fuel was cheap. By 1733 there were seventy or eighty Newcomen engines in Britain—about six in Cornwall, fifty at coal mines in the Midlands and North, and others elsewhere.[5] By the 1730s also, Abraham Darby's coke-fired blast furnaces, first used in 1709, had helped to establish Shropshire as a major iron-producing region.

The river navigation that linked Derby with the Trent was surveyed by an engineer named George Sorocold, a central figure in other improvements of the period. He was one of the engineers for the construction of Liverpool's first dock (c. 1709–20), and before that had gained a reputation for the installation of piped-water systems in many provincial towns. Like similar water-supply schemes in Paris, Sorocold's installations used waterwheels to pump water out of rivers into a cistern, from where it was distributed to the basement kitchens of individual houses.[6] Towns where Sorocold installed waterworks of this kind included Derby, Leeds, Norwich, Portsmouth, Sheffield, and Newcastle in the 1690s, with the famous London Bridge waterworks following in 1701, and a smaller scheme at Bridgenorth in 1702. His partner in many of these projects was John Hadley, who in 1699 became engineer for the scheme to make the River Aire navigable to Leeds.

These water-supply schemes were usually operated by private companies and at this time served only small parts of the larger towns. In London, apart from the London Bridge waterworks, there was the New River Company that brought water into the town from Hertfordshire by means of a canal constructed in the seventeenth century, and yet another company had a Newcomen-type steam engine at York Buildings near the Strand. The latter engine served houses in an area that not long before had been mostly green fields separating the cities of London and Westminster. Lincolns Inn Fields and Covent Garden had been built up in early seventeenth-century town-planning schemes, but now dignified squares and streets were extending farther to the north and west, around the new churches of St George's, Bloomsbury (1720), and St.-Martin-in-the-Fields (1721). Previous to that, the orderly rebuilding of the city after the Great Fire of 1666, and the completion of St. Paul's around 1710, had likewise added to the Londoner's sense of prosperity and well-being.

But a venture that was less in the public eye, though more significant for the future, was a water-powered silk-twisting mill at Derby. Sorocold had built a waterwheel for it in 1702, and then in 1717 he was involved in extending and adding to it. By this time it belonged to John and Thomas Lombe, who had experienced the early difficulties of the scheme and had traveled to Italy to look at the silk-twisting mills being used there. By 1721 the Derby silk mill was a success, and soon there were several similar establishments in Derbyshire and Staffordshire.

Much of what took place during this period of rapid technical change must be understood in terms of the discussions about technology that took place in England during the 1640s and 1650s. In the next decade, too, lofty objectives concerning the improvement of trades and manufactures were discussed in the Royal Society, but whether this body provided any real stimulus to the practical arts is difficult to assess. However, the discussions of the Civil War period definitely had some influence. For example, agriculture was a particular interest of Samuel Hartlib; while he was encouraging its study in England, Royalists fleeing from Cromwell's rule traveled the continent, observing agricultural techniques and thinking of applying them to their own lands at home. A major problem in English agriculture earlier in the seventeenth century had been that a lack of winter feed severely limited the number of sheep and cattle that could be kept. Travelers in the Low Countries saw clover and other legumes being sown to provide fodder crops, and also improved strains of grass, such as cocksfoot, rye grass, and meadow fescue.

Clover made good hay for animals and added to the fertility of the soil (by nitrogen fixation); the increased herds of cattle produced more manure and so provided a more effective recycling of organic material. The theory of all this was not understood, but the improvements to both soil fertility and cattle rearing were readily appreciated and proved to be particularly effective in East Anglia. In the poorer hill-farming areas of northern England, evidence that more fodder was being produced and more cattle were being overwintered is still evident in the numerous hay barns built during the seventeenth and especially the eighteenth century, each with stalls at one end where six or eight cows could be kept during the winter and where storage for hay extended to a loft over the cows. In some places these buildings were for hay only, but in areas where cereal crops were also grown, barns would also have space for storing these and a central threshing floor (figure 26).

Figure 26
Fodder crops in the seventeenth and eighteenth centuries, including (left to right) cocksfoot, meadow fescue, ryegrass, sainfoin, and two kinds of clover. Although these plants were grown mostly in the English Midlands and East Anglia, there is evidence of increased fodder production in northern England also. Thus many barns were enlarged or newly built so that more hay could be stored and more cattle over-wintered. The example shown was built in 1776 and has storage for hay and cereal crops, also a threshing floor (large entrance arch), and accommocation for cattle (small doorway).

Among those who had introduced the new fodder crops into England, notably by importing Dutch clover seed, the best known was Sir Richard Weston, who had been in the Low Countries in 1644. Hartlib's interest had been immediately aroused, and he too made inquiries about the continental fodder crops and helped to spread knowledge of them.

This was a period when the English were learning fast from apparently less backward European countries—from Italy about silk twisting, from France about canals, and from the Low Countries about agriculture. There were also the locally invented steam engines and iron-making processes. And the net result was that already, in the first decades of the eighteenth century, England was probably the wealthiest country in Europe in relation to its size, although France, with three times the population, was more powerful.

The Mood of the Improvers in England

The authors of travel diaries in England at this time, such as Celia Fiennes (1698) and Daniel Defoe (1720s), recorded a wide range of industrial activity, including coal mining in Yorkshire and the silk mill at Derby.[7] But river navigation, water-supply projects, and civic improvements—new houses, roads, and churches—probably attracted the most general attention. The progress being made was obvious and visible, but there was also much cause for dissatisfaction. In the 1730s, roads were still very bad throughout most of the country, and there were a number of delays about building the new bridge over the Thames at Westminster, which was eventually begun in 1739. So there was a considerable need for public works of many kinds, and the lack of skilled men in the relevant branches of engineering was becoming obvious. These points were alluded to by Alexander Pope in 1731, in a poem addressed to the Earl of Burlington. Its closing lines give a strong impression of the sense of prosperity and well-being that the construction of new roads, churches, and bridges would bring.

Bid Harbours open, public Ways extend,
Bid Temples, worthier of the God, ascend;
Bid the broad Arch the dang'rous Flood contain,
The Mole projected break the roaring Main;
Back to his bounds their subject Sea command,

And roll obedient Rivers thro' the Land;
These Honours, Peace to happy Britain brings,
These are Imperial Works, and worthy [of] Kings.[8]

Pope wished to stress the benefit to the community arising from public works such as these, and elsewhere he picked out for commendation men who used their own resources to promote projects for the public good. A prime example was John Kyrle of Ross-on-Wye,[9] who, around 1700, had provided his home town with a water supply on the Sorocold model, and also a public park. The two things were connected, for the water pumped from the river by the water wheel was used to work fountains in the park before it was piped to various points in the town for domestic consumption.

Another man whom Pope admired was his friend Ralph Allen, improver of postal services, owner of quarries near Bath, and one of those responsible for the Avon navigation in the vicinity of Bath. Allen made a fortune from these activities, but his private gain was the product of public improvement, and he used his money for many charitable purposes.[10] From 1725, Allen actively exploited building stone from his quarries for the construction of the famous squares and crescents of Bath. The quarries were celebrated in the technical literature of the time. partly on account of a railroad or wagonway with wooden rails by which the stone was carried from the quarry face to a wharf on the River Avon. The wagons on this railroad had flanged wheels with good brakes, which were necessary because of the steep hillside down which loaded wagons descended to the wharf by gravity, empty ones being hauled back to the quarry by the use of horses. Specially designed cranes lifted the stone from the wagons onto boats, which carried it a short distance down river to a point near the building sites in Bath. In the eighteenth century this was thought to be a "great Improvement on some Carriages and Waggonways made use of at the Coal Mines near Newcastle."[11] It was also celebrated in verse—though not Pope's—as the

. . . *New Made Road,* and wonderful *Machine,*
Self-moving downward from the Mountain's height.[12]

Alexander Pope's rather general comments on such things are relevant here because he had many friends among public men in England, including some like Allen, who commissioned public works or in other ways contributed to the material progress of the nation. To some extent, then, his poems express the aims of these people and describe social ideals that were capable of influencing engineering projects. Englishmen at this time were prone to use the language of the Renaissance to discuss such ideals. They even, in their most confident moments, talked about a golden age that had just dawned or was about to dawn, and described its qualities in terms of the cultivated grandeur and noble civilization of the Roman Empire. The achievements of ancient Rome, they thought, were being equaled and could be surpassed in modern times.

What Pope was advocating was exactly what some Italians of the period of Palladio and Galileo had done (chapter 3). They too wished to emulate the splendors of ancient Rome and concentrated much of their effort on canal and river works or civic improvements. Their leading technological skills were those concerned with machines, hydraulics, architecture, and public works of all kinds.

Pope would not have been displeased by such a comparison with sixteenth-century Italy, for he linked the name of Palladio with that of Vitruvius, the Roman architect. Palladio had provided a fine example of the emulation of Roman architecture in Venice, equaling what Vitruvius and other architects had done for Augustan Rome. And partly because it was in building towns, roads, harbors, and water-supply systems that the Romans had excelled, not in manufactures or metallurgy, it was works of the former kind that were most often stressed in Renaissance Italy, in Colbert's France, and now in England.

English country gentlemen with their large estates were also often conscious of the need for agricultural improvement. Here Palladio was a more appropriate ideal for them than perhaps they realized, for it will be recalled that many of his houses were built for gentlemen farmers who were developing newly reclaimed land around Venice. Probably Pope did not know much about the agricultural background to Palladio's architecture, but he did criticize gentlemen who surrounded their country houses with vast lawns, ornamental gardens, fountains, and avenues. For him, the ideal country gentleman was one

Whose ample Lawns are not asham'd to feed
The milky heifer and deserving steed;

Whose rising Forests, not for pride or show,
But future Buildings, future Navies, grow . . . [13]

As the eighteenth century progressed, more landowners were moved by similar sentiments, and steady progress was achieved on the basis of the agricultural improvements introduced in the later seventeenth century. More fodder crops such as clover were grown, more hay was harvested, and more animals were kept. On the village level, small farmers gradually abandoned, or were forced to abandon, traditional methods of farming small strips on open fields. Land was enclosed, and larger, more efficient farms were formed.

In many ways England at this time was catching up with what had been done on the continent many years before; but besides catching up, the English were taking new initiatives and outpacing their European rivals. Among many English people there was a marked sense of optimism and confidence. It was felt that the glories of ancient Rome were being recaptured and surpassed. To do this had long been the aim of Italians and Frenchmen as well, and the literary notion of Augustanism was merely a latter-day expression of a Renaissance sentiment. One important difference, however, between the way in which the English put these things into practice as compared with the French, or earlier with the Venetians, was that in England matters were usually left to the initiative of individuals or small groups of men, and government was relatively inactive.

Fifty years before Pope, Colbert had certainly thought that harbors, roads and bridges were "Imperial Works," and had organized their construction under government supervision. When Colbert promoted such schemes, he was thinking not only of the economic development of France but also of the imperial splendor of Louis XIV. Indeed, Colbert's aim in his economic policies has been characterized as "to glorify the King" and to create "the grand style that was to signalise the reign of Louis XIV." Thus Colbert tolerated the grandiose building scheme at the Palace of Versailles, which drained away much of the wealth he had so carefully husbanded. The Languedoc Canal, too, was intended to be a monument to the king as well as an economic improvement. "Take care that your works are constructed in such a way that they will last for ever,"[14] Colbert commanded, no doubt recalling the Roman aqueducts that indeed seemed capable of lasting forever.

Mercantilist Economics and the Eastern Trade

The imperial splendor about which seventeenth- and eighteenth-century Europeans talked was not conceived on an international scale, but in terms of individual nations. Pope wrote:

Oh when shall Britain, conscious of her claim,
Stand emulous of Greek and Roman fame?[15]

This nationalist emphasis had also been characteristic of Colbert. In his time he was the outstanding exponent of the economic theory known as "mercantilism," and this has been described as "the economic phase of nascent nationalism."[16]

Mercantilism influenced the policies toward technology adopted by several European states during the eighteenth century. It was a theory that pictured trade as a savagely competitive struggle between nations. Wealth was defined in terms of the possession of gold and silver bullion, and unless a nation owned gold or silver mines, it could only become wealthy by trading at the expense of others. The French could increase their prosperity only by reducing that of the Dutch, because if the bullion reserves of France increased it would most likely be because French merchants traded to the disadvantage of Dutch colleagues. Prosperity depended on having a favorable balance of payments, so that gold flowed into the Treasury. But because it was impossible for every nation to achieve this at once, the gains of one must come from the losses of others. A nation pursuing mercantilist policies would aim for self-sufficiency in its industries, so that no imports were necessary. Then foreign trade need be pursued only in circumstances where it led to a net gain in the nation's bullion.

This view of economic life was given a peculiar twist in the eighteenth century by the increasing volume of trade between Europe and the Far East. Spices grown in the East Indies, cotton goods from India, and silk, porcelain, and tea from China loomed large in the import trade of most European countries. At this time, "India was the greatest exporter of textiles the world had ever known, and her fabrics penetrated almost every market of the civilized world. The extent to which Western Europe shared in this trade is reflected in language itself: "chintz," "calicoe," "dungaree," "gingham," "khaki," "pyjama," "sash," "seer-sucker," and "shawl" are all Indian words."[17]

During the seventeenth century, the majority of these goods were carried to Europe in Dutch ships. The Dutch had been active in this trade since the end of the sixteenth century, and a number of trading companies were grouped together under government auspices in 1602 to form the Dutch East India Company. The English also had an East India Company founded at about the same time, but although it operated with some success it was not at first so large or powerful as the Dutch concern.

Trade with the East was a somewhat dubious undertaking from a mercantilist point of view, because there was very little that Europeans could export to India and China. Most of the porcelain, silk, or cotton had to be paid for in gold or silver. The quality of European textiles and pottery was markedly inferior to what was available in the East, so it is understandable that European wares of this kind could not be exported in any quantity.

Thus while wealthy people in Europe bought Chinese and Indian goods in large quantities because of their high quality, there was no corresponding demand on any large scale for European goods in the East. For that reason, European bullion was steadily drained away to India and to the treasury of the Chinese Empire, and it was only by using the gold and silver of Central and South America that European nations were able to maintain this trade. Another difficulty was that European textile manufactures were suffering as a result of competition from Indian products, and the import of cotton cloth from India was prohibited in France in 1686 and in England in 1701.

However, these prohibitions were short-term measures. The proper mercantilist solution to this imbalance in trade was to promote the development of industries in Europe that could produce goods of the same quality as those from the East. With regard to textiles this object was not secured until the nineteenth century, when competition from the British cotton industry at last virtually destroyed the textile trade of India. But the problem of producing something comparable with Chinese porcelain was solved earlier in the eighteenth century in Saxony, largely as a result of researches provoked by mercantilist policies.

Chinese porcelain began to come to Europe in significant quantities during the 1640s. Fairly soon, Europeans were investigating the technique of making porcelain, which was so much finer and whiter than their traditional earthenware. There were problems about the kinds of raw materials to be used, and problems too about the very high temperatures at which the pottery was fired. By a long series of experiments with materials that seemed likely to give the required white pottery, and by trial and error with various firing temperatures, a satisfactory imitation of Chinese porcelain was eventually produced at Dresden in 1708. Tschirnhausen, the man mainly responsible for this breakthrough, set up a pottery at Meissen in Saxony.

The importance of trade with the East for European technology during this period was that techniques and processes unfamiliar in the West provided a very considerable stimulus to Western inventors and experimenters. Historians often mention the debt Europe owed to China in the twelfth and thirteenth centuries for such things as paper and the magnetic compass. Few have recognized, however, that the seventeenth and eighteenth centuries mark a second phase of Chinese and Indian influence on European technology. It was not, as in the earlier period, that the technologies of the East were always more advanced than those of Europe, but merely that they were different. And difference, especially in quality, suggested new products and new standards of manufacture to inventive European minds. This important "stimulus effect," and its role in the eighteenth century, is discussed at length in a separate volume.[18]

Mercantilist Policies in Europe

Mercantilist thinkers were much troubled by the imports coming into Europe from the East and by the loss of bullion that this entailed. Like Colbert, they saw a major role for government in any response to this threat—a response that would include improvements in transport, mines, and manufactures in their own countries. French governments had a long tradition of intervening directly to promote such improvements, going back before Colbert's reforms.

British traditions were very different, however. During an earlier period of economic expansion, between 1540 and 1640, English governments had attempted to manage the development of mines (chapter 5) and to control other types of industrial enterprise, but not very effectively, and with some entrepreneurs in the industries concerned evading government regulation. Thus already in the early seventeenth century there was a significant contrast between the prominence of privately run enterprises in England and major government-directed industries in France.[19] Under these conditions, the English economy had grown faster, but the French were more successful in manufacturing luxury goods, such as silks and tapestries, and were beginning to develop greater competence in engineering.

After 1640 events had tended to accentuate this contrast. The English had fought against the absolute rule of kings, and even after restoration of the monarchy in 1660, its powers were limited. At the same time, the "Protestant ethic" had emerged in its most individualistic form, and the idea was gaining ground that prosperity was best promoted by giving individuals freedom to pursue their own economic self-interest. There was a clear and sharp contrast here with the mercantilist tendency to believe that "the state could, should, and would" act in the economic sphere.

In France, Colbert thought that the state should manage difficult or risky ventures, such as the Gobelin tapestry factory or the French East India Company (founded in 1664). He also thought that state enterprise was usually preferable when it came to the construction of roads, canals, and other public works. But he believed that ordinary manufactures and commerce were best left in private hands, though governments might still take action to encourage or regulate them. One idea was that the *Académie Royale des Sciences* should collect and circulate practical information on trades and industries. Many years later, from 1761 onward, this led to the publication of a series of profusely illustrated "descriptions of the arts and crafts."

A similar project, though carried out by men who favored a free economy and were opposed to mercantilism, was the great thirty-five-volume encyclopedia edited by Denis Diderot and Jean d'Alembert and published between 1751 and 1780, the *Encylopédie . . . des sciences, des arts et des metiers*. This referred several times to the English author Francis Bacon and was clearly inspired in part by Bacon's views on how knowledge of trades should be searched out and organized. Diderot showed particular enthusiasm for visiting factories and workshops to collect material, and his comments in the encyclopedia entry on "Art" show his awareness of ideas about the division of labor and his knowledge of Galileo's ideas concerning the size and strength of machines.

Among the industries that French governments tried to regulate was the dyeing of textiles. One negative result was that many French dyers moved to Switzerland, but a more positive outcome was the appointment of a series of experts on chemistry to the position of director or inspector of dyeworks. Through their researches, new dyeing methods introduced from India and Turkey were a little better understood, and then in 1785, chlorine bleaching was developed by C. L. Berthollet (who had been made inspector of dyeworks the previous year).[20]

Similarly, government control of road-building led to the accumulation of greater expertise through the gradual development of a professional corps of civil engineers, the *Corps des Ponts et Chaussées*. And just as Colbert's improvements to the French navy had demanded the development of a whole range of schools for ship building, hydrography, and artillery, so now the government organization for road building called into being, from 1747, a school for the training of civil engineers. The result was that during the eighteenth and much of the nineteenth century, France had the best-trained and often the most competent engineers in the world.

A century after Colbert, an interesting exposition of mercantilist theory was given in a masterly textbook on mining published in Vienna in 1773, with a French translation following almost immediately. Its author was Christoph Traugott Delius (1728–79), and it provided a comprehensive and detailed account of the mining arts: sinking shafts, cutting rock, mine ventilation, lifting gear, pumps, and the treatment of ores. The discussion of mercantilist economic theory cropped up in a discussion of the finance of mines.[21] The prosperity of a nation, Delius said, was directly dependent on the amount of money circulating within it. If each country produced all the goods necessary for its welfare, including luxury articles, then its prosperity would neither grow nor decrease. But if a country was able to produce goods or materials that other countries needed, then it could trade favorably with its neighbors. Bullion would flow into its treasury, and its wealth and general prosperity would increase.

Delius wrote his book while teaching at the Schemnitz Mining Academy, situated in the Hungarian lands of the Austrian Empire. He remarked that Austria had a considerable advantage over many other nations, for within its empire several mines produced silver and a little gold. Thus by regulating not only its trade but also its mines, Austria could readily control its "balance of commerce," and ensure that its bullion reserves, and so also its wealth, were always increasing. Between 1740 and 1772, the mines of Hungary and Transylvania had provided the Royal Treasury with gold and silver worth a total of 150,000,000 florins, equivalent to about 60,000,000 ounces of silver. A measure of the significance of this sum is that it seems to have provided something like 10 percent of the total revenue of the Austrian Crown during the period.[22] But even with this large supply of bullion, the imports of the Austrian Empire could only just be paid for. Until 1775, when exports at last exceeded imports, there was a chronic imbalance of trade, and most of the silver from Hungary was used to make up the deficit. Thus, Delius explained, there was an urgent need to establish new industries within the empire so that imports could be cut. There were some people, he said, who disputed the value of mining and who cited Holland and France as nations that had become rich while lacking mineral resources. But these nations had no certain source of wealth and had become wealthy only through the weakness and idleness of others. If neighboring countries developed their own industries more effectively, he argued, they could become less dependent on France and Holland, and keep their own wealth within their boundaries. It was assumed in this argument that if Holland failed to sell its imports from the Indies, if France could not sell its wine, and if neither could sell their cloth and other manufactures, then they would be as nothing.

The weakness of mercantilist theory is apparent. There was a failure to see that the growth of manufacturing or agricultural production could result in an increase in wealth, even if it was all consumed within the country and there was no inflow of gold. Gold and silver constitute wealth only if there are products that they will buy. If manufactures increase, it can come about that the same amount of bullion will buy more goods, or that more goods are bought on credit. Thus the wealth of nations can increase by their industry and not only by trade conducted to the disadvantage of others.

Because of the special importance of gold- and silver-mining in the mercantilist scheme of things, it is not surprising to find that like road and canal building in France, mining in the Austrian Empire was largely a government enterprise. There was provision for the private ownership of mines, but if the production of a private mine declined, it could be taken over by the Crown.

The state-owned mines were administered by professional civil servants who needed to combine some technical knowledge and practical skill with the ability to keep the detailed written records required by a distant government administration. At first the officials who ran the mines were trained by a system of apprenticeship, but in 1733 attempts were made to organize some part-time formal education for trainee officials. Courses of this kind were begun in 1735 at Schemnitz[23] in Hungary, the center of one of the most important mining areas within the Austrian Empire. In 1763 a more ambitious school of mining was established at Schemnitz, and in 1770 this was reorganized as the Royal Hungarian Mining Academy. For a few years it had a high reputation throughout Europe as a kind of technological university. Its courses in practical chemistry were particularly well known, and in 1770–72, while Delius was there, the course on mine machinery must also have been very good. But in many ways the kind of education provided at Schemnitz had been anticipated by the small technical schools founded in France from Colbert's time onward. The military schools at Mézière (founded 1749) and at La Fère (1720) were probably the best academically, but, more to the point, the *École des Ponts et Chaussées* had begun teaching mining engineering on a small scale in the 1750s. There were also mining schools elsewhere, notably at Freiberg in Saxony, where the *Bergakademie* was founded in 1765. This institution was comparable with the academy at Schemnitz, but had the advantage of being less isolated. It was not far from either of the two universities in Saxony: Wittenberg and Leipzig; and the geology lectures given at Freiberg at the end of the century by A. G. Werner attracted students from far and wide.

The mining districts of Saxony figured prominently in the previous chapter because of their association with Agricola. Since then, mining, like other local industries, had been badly disrupted by the Thirty Years' War (1618–48), but at the beginning of the eighteenth century, silver output in Saxony had regained its prewar level. As the century proceeded, Saxon output steadily rose, while from about 1780 the much larger output of the Schemnitz mines began to fall (figure 27). The mining area in Saxony was centered on Freiberg; at the end of

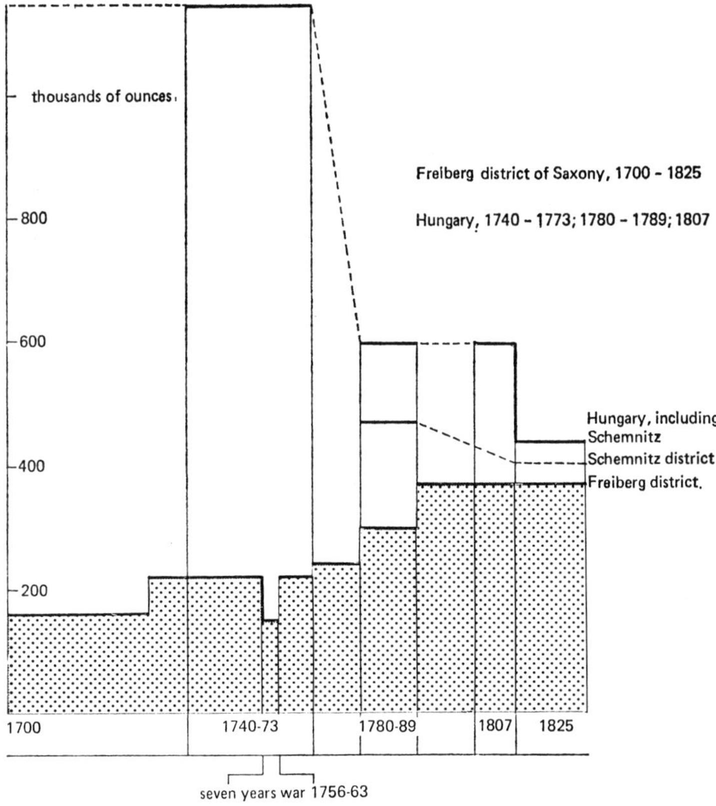

Figure 27
Average yearly output of silver from the Freiberg district of Saxony and from the Hungarian mines.

(Based on figures compiled by D. M. Farrar using data from *Bergakademie Freiberg Festschrift 1765–1965*, Freiberg, 1965, and from C. T. Delius, *Anleitung zu der Berghaukunst*, Vienna, 1773, also Héron de Villefosse, *De la richesse minérale*, Paris, 1819.)

the century, production there was about 400,000 ounces of silver per year. This was about the same as the output at Schemnitz.

This seems to reflect a somewhat healthier state of economic affairs in Saxony than in the Austrian Empire. Saxony was a country of about 2,000,000 people which, by 1800, had many of the characteristics of an industrial region. Between 1756 and 1763, it was again badly affected by war, but a government commission for reconstruction helped to establish seventy new industrial enterprises in the twenty years up to 1784. Coal was mined on a scale that was still very unusual in central Europe, and production was 60,000 tons in 1807. There was a thriving textile industry and the famous Meissen china factory as well as mines that produced copper, silver, lead, and iron.

Mercantilist thinking predominated in the policies of the Saxon government. The mines were subject to close government control. The academy at Freiberg, like that at Schemnitz and like the technical schools in France, was financed by the government and was founded, like the other schools, to provide technically qualified personnel for government service. These educational institutions were therefore a direct outcome of the mercantilist policy of state management in the mines and in public works engineering.

Steam Power in the Austrian Empire

The use of atmospheric steam engines in the Austrian Empire not long after their first application to English mines indicates how well the practical arts fared under the influence of mercantilist policies. By 1735 there were already at least three engines in the Schemnitz area: two at the Windschacht mine and one not far away at Königsberg. Two further engines seem to have been installed at Windschacht in the next two or three years.

These engines, employed to pump water from the mines, were built by the Austrian architect-engineer Joseph Emmanuel Fischer von Erlach (1693–1742); he had visited England, studied Newcomen engines there, and engaged a certain Isaac Potter to return to Hungary with him. Potter helped particularly with the Königsberg engine, but died in 1735, before Fischer's later engines were begun.

A few years later, a further atmospheric steam engine was erected by Joseph Karl Hell (1718–85), an engineer trained at Schemnitz in the 1740s. Hell pioneered the use of water pressure to drive a hydraulic piston engine, again for working pumps. He also invented a most unusual pneumatic engine for raising water from the mines. Both types of engine were powered by water supplied from a high-level reservoir, and nine of the hydraulic engines were built between 1749 and 1768.

Other Newcomen engines on the continent were few and far between (figure 28). There was one in the coal- and iron-working district around Liège, near Brussels, and one had been built at a Swedish mine by Marten Triewald, another foreign engineer who had visited England. In Saxony, people had been interested in the use of steam power for pumping out mines even before Newcomen's engine had proved to be successful, although no Newcomen engines were used there until much later. In 1690 Denis Papin, Huygens's former assistant, wrote to Count Philip Ludwig von Zinzendorff, assuring him that a steam engine would be "very advantageous" for pumping out mines. Zinzendorff was an influential man in Saxony, and had written to Papin inviting him to visit a mine in Bohemia which had become unworkable owing to underground flooding.

Two books published in Saxony in the 1720s show that there was continuing interest in the subject. One was written by Jacob

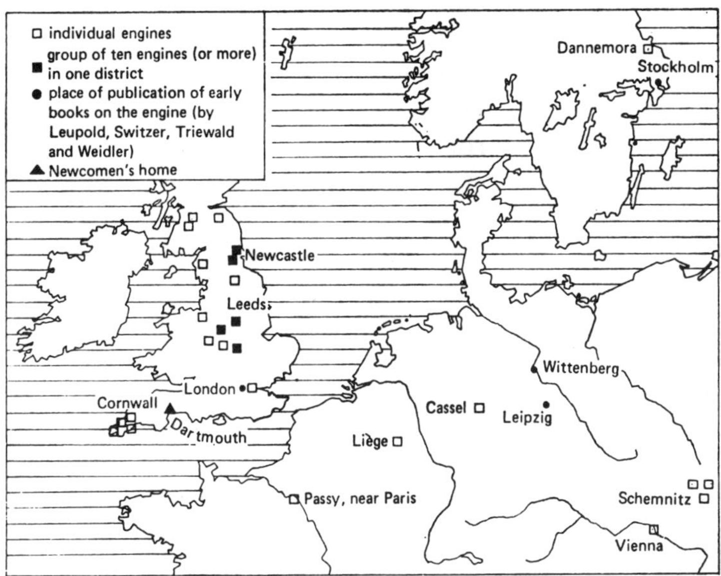

Figure 28
The Newcomen engine in Europe up to 1734.

Leupold, a Leipzig instrument maker and mathematics teacher who in 1725 became a commissioner of mines in Saxony. In the first volume of a remarkable encyclopedia or "theater" of engineering, which appeared in twelve volumes from 1724 onward, Leupold described the Newcomen-type engine that Isaac Potter had helped to erect near Schemnitz, and quoted letters from his "good friend" Fischer von Erlach. Another book, a "treatise on hydraulic machines" that was written in Saxony by Johann Friedrich Weidler, a professor at Wittenberg University, gave more details of early Newcomen engines. It illustrated one erected at York Buildings, London, in 1726 that was designed to pump water for the supply of houses in the Strand.

Weidler wrote in Latin, thereby setting himself the intriguing task of devising Latin technical terms for the components of Newcomen's engine, the "English machine." He said that the first such engine in Germany was erected at Cassel in 1722 by Fischer von Erlach. Cassel had been a place where much work had been done on steam engines since the time, around 1700, when Papin lived there, but it is doubtful whether a Newcomen engine *was* ever erected there. However, it is quite possible that Fischer von Erlach visited Cassel on his way back to Austria, and he may have provided a small engine or model for Landgrave Charles, ruler of Hesse-Cassel and former patron of Papin.

Apart from this, and apart from the mine engines near Schemnitz, Fischer von Erlach was responsible for a steam engine erected in Vienna in 1727. This pumped water needed to work fountains in a garden and was commissioned by Prince Francis Adam of Schwarzenberg. It had a two-foot (0.6 meter) diameter cylinder, and was notable chiefly for a novel boiler designed by Fischer. "The round copper boiler was surrounded by many closely spaced cells," Weidler wrote. The smoke from the fire passed through these cells before going up the chimney; it "went round the boiler twice, was passed back and forth, and any heat remaining in the smoke, even that was drawn out of it."[24]

Directions of Technical Progress under Mercantilism

The fact that this novel engine in Vienna was built in connection with an ornamental garden and not for an industrial purpose helps to explain why economic and technical development in the Austrian Empire was lagging by the end of the eighteenth century. Magnificent palaces and gardens were built by aristocrats, and very advanced technology was applied to the particular tasks in which the government and the nobility were interested. But much of the countryside remained very backward, and basic industries like textiles were underdeveloped. The vast sums that were spent on imperial magnificence in Vienna, and the expenditure on mine machinery and on the academy at Schemnitz, probably drained resources away from more mundane but more urgent tasks and certainly contributed to Austria's chronic balance of payments difficulties. Delius said that because of the demand for food and other materials it created, the mining community at Schemnitz had a beneficial influence on the agriculture of the surrounding area. But his comments on the difficulty Austria faced in maintaining its bullion reserves, despite its rich silver mines, suggest that agriculture and manufacturing production was really very inadequate.

The situation in most of the territory ruled by Austria was in some ways similar to that which is found in many underdeveloped countries in the modern world, which economists call a "dual economy." That is, advanced technology and good technical education are to be found in a small sector of the economy, with a few people benefiting, while the majority of the population live mainly by agriculture, using traditional techniques with low productivity. Austria used Hungary as a source of food and raw materials and allowed the industries of Hungary to decline. Only in the mining towns was there a technically advanced industry, and this had stronger links with Vienna and western Europe than with the immediate needs of people in the surrounding area.

Backward conditions in agriculture or industry were of course by no means unusual in eighteenth-century Europe, but in some countries they were less accentuated. In England, less power and money was concentrated in the hands of the monarch than was the case with the Austrian Emperor, and in English society generally, wealth was more widely diffused among members of a fairly large middle class, many of whom were active in trade and industry. And while large-scale projects as ambitious as the Languedoc Canal or the Schemnitz mines were scarcely attempted in England before 1750, a great number of more modest schemes provided a larger volume of work for craftsmen and engineers and ensured a more balanced development of the country as a whole.

The effect on technology can be illustrated by comparing the Schemnitz mine engineer J. K. Hell (1718–85) with his English contemporary, John Smeaton (1724–92). In their work on Newcomen engines and horse gins for mines and in their interest in piston engines worked by water pressure, Hell and Smeaton were engineers of very similar caliber. But while in Hungary there was little an engineer like Hell could turn to outside the mines, in England John Smeaton drew plans for harbors, bridges, canals, and land-drainage works, and the half-dozen mine engines he designed occupied only a fraction of his time. By the end of the eighteenth century, the Austrian Empire's shaky finances seem to have inhibited investment at the mines to the point where maintenance was neglected. Thus while the young men who worked under Smeaton became the civil engineers of a new generation, Hell's pupils were faced with a declining industry and few openings for engineering work elsewhere. So the brilliance of engineering at Schemnitz in the time of Fischer, Potter, Hell, and Delius was not developed further.

It is easy to criticize this state of affairs, but it should be remembered that eighteenth-century governments did not envisage their objective as economic development in the modern sense. Their aim was the acquisition of wealth and power; but more than that, they wished to express the achievement of their civilization in a display of imperial magnificence that would rival the grandeur of Rome—and outshine the Empire of China, of which they were newly conscious. Thus the prudent and austere Colbert, who did so much to build up the wealth of France, did not object to the expenditure of 50,000,000 livres[25] on the palace, gardens, and water supply at Versailles, even though this strained the country's economy. For what was the object of possessing wealth if not for a display such as that?

In Austria, the royal house and the nobility also built great houses, churches, gardens, and palaces. An example is the palace of the Prince of Schwarzenberg, where steam-powered fountains were installed in the 1720s, as mentioned above. These were designed by Fischer von Erlach, whose father had been the architect of the palace itself, which was built between 1705 and 1720. Every minor princeling wanted his own Versailles, and the two Fischers did much to satisfy their demand. Indeed, the elder one nicely expressed his aspirations by publishing a book of pictures "in the representation of the most noted buildings of foreign nations, both ancient and modern."[26] The palaces and bridges of contemporary China were shown along with those of ancient Rome, Greece, and Egypt. Then palaces and churches in Vienna were illustrated, all of them designed by the author, and the reader could not but feel that these were the equal of anything that other civilizations had achieved.

The English shared the ideal of emulating the splendor of ancient Rome, but interpreted it with a somewhat utilitarian emphasis. Roman grandeur was the result of Roman skill in building roads and aqueducts. Magnificent architecture was not just an empty show, but the product of wise government and a concern for the public good. As Alexander Pope put it:

'Tis Use alone that sanctifies Expence,
And Splendour borrows all her rays from Sense.[27]

And as another poet, the Reverend John Dalton, explained in 1755, it seemed right to celebrate the growing prosperity of England in the same kind of language as Roman poets had used in describing the achievements of their own civilization. Thus Dalton employed "classical allusion" and "parody" of Roman poets to describe the progress of agriculture, mining, and forestry in his native Cumberland and to portray a Newcomen engine at Whitehaven (which he wrongly attributed to Savery).

Sagacious Savery! Taught by thee
Discordant elements agree,
Fire, water, air, heat, cold unite,
And listed in one service fight,
Pure streams to thirsty cities send,
Or deepest mines from floods defend.
Man's richest gift thy work will shine;
Rome's aqueducts were poor to thine![28]

In these ways, then, the English had their own ideas about the emulation of Rome. Landowners were enthusiastic "improvers" as often as they were palace builders. The King was not an absolute monarch and rarely indulged in such extravagant royal expenditures as those of the Austrian or French rulers. Parliament set limits on the King's powers and supported a form of government in which policies like Colbert's, or like those of the Austrian Crown, were almost unthinkable. One disadvantage of this was that technical schools like those of France, Saxony, and the Austrian Empire were never established in Britain. In the nineteenth century, this neglect of technical education came to be sorely felt. But in the eighteenth century, Britain gained by avoiding the defects of mercantilist policies.

Not only did Britain differ from its continental neighbors with regard to the role of government. There were also differences of view within Britain, particularly about how resources and wealth should be used. The emphasis of Alexander Pope and the other poets quoted was on the use of wealth for the public good, that is, for improving roads and harbors, erecting public buildings, and so on. Pope praised men who invested in projects of this sort, or in works of charity, or in tree planting and "husbandry." To use wealth in these ways was the opposite of the individualistic application of resources solely to promote one's economic self-interest, and indeed, Pope criticized men who ran their estates on a wholly capitalistic basis, reinvesting profits entirely in developments that would make them individually richer regardless of public benefit.

There is thus an interesting parallel between what Pope was saying about the use of wealth and what Francis Bacon wrote about the use of knowledge a century earlier. In both authors there was a threefold stress on public benefit, good husbandry, and works of charity. The first category could encompass architecture, especially public buildings, as well as roads, bridges, and water supplies—everything the French included under the heading of *architecture hydraulique*. The second referred to agricultural improvement and tree planting, and the third could embrace medicine as well as the giving of alms.

The significant contrast, which may become clearer in the next chapter, is between these *social ideals* for the use of knowledge, wealth, and hence technology and the increasingly prevalent attitude in England that the public good was best served if individuals pursued their own economic self-interest. This attitude was sanctioned by the version of the Protestant ethic that emerged only after Bacon's time. It was fostered by the belief in economic freedom that characterized Britain as compared with the continent. And it was soon to be enshrined in economic theory by Adam Smith.

7

Technology in the Industrial Revolution

Steam, Coal, and Iron

When Newcomen's atmospheric steam engine was first introduced into the mining areas of Britain, horses were the most common source of power in the mines. They hauled wagons of coal along short lengths of railroad; or, walking in a circle around a horse engine or "gin," turned winding gear and worked pumps. So when a mine owner thought of installing a steam engine to drive pumps in his mine, he needed to compare its promised performance with what his horses were already doing.

One such comparison was made in northeast England during 1752, at a coal mine 240 feet (72 meters) deep. It cost 24 shillings a day to keep horses for driving the pumps. They worked two at a time, in eight three-hour shifts, and they lifted 67,200 gallons (305,000 liters) in a full day. The steam engine, on the other hand, cost only 20 shillings a day, and in that time it could pump *four times* as much water from the mine.[1]

Two sources of energy were involved here, namely fodder for horses and coal for the engine. It was characteristic of most industrial activity at this time that it depended on land-based resources, that is, on the products of agriculture (such as fodder). There were many competing demands on the land for food production, fodder, textile fibers (linen and wool), timber for construction, firewood, and wood for charcoal making. As the previous chapter indicated, improvements in agriculture had been under way for some long time and had contributed to increased output of most land-based products—not only food but also most notably fodder. However, one other major trend was the increasing use of coal as a substitute for some land-based energy sources.

This trend was more evident in Britain than in almost any other country, with the expansion of coal mining in northeast England and parts of the Midlands already noticeable before 1600. During the seventeenth century, the price of firewood for those who had to buy it increased faster than most other prices, encouraging changes in domestic fireplaces and chimneys so that coal could be used as a household fuel. The same economic pressures encouraged the use of coal or coke as a substitute for firewood or charcoal in a wide range of industries, including glassmaking, brewing, brickmaking, and metalworking. There were brass foundries in Bristol that made the transition to coke around 1700. The biggest challenge, however, was to find a way of using coal in iron smelting, something that several people tried, though without any success until 1709.

In existing iron-smelting areas in Britain, such as Kent or the Forest of Dean, the production of charcoal for the furnaces by "charring" small timbers in closed heaps was a major subsidiary industry. This fuel supply could be sustained indefinitely by coppicing trees and then letting them regrow. After as little as ten years, young growth from the stumps would be worth harvesting again for charcoal making. Thus by using different parts of a fixed area of woodland in rotation, a continuous supply of charcoal was obtained, and there was no shortage of fuel for established furnaces. However, in a country that needed land to meet increasing demands for food, fodder, timber, and textile fibers, it was becoming impossible to develop woodlands to support new furnaces in an expanding industry.

One man with experience of the fuel problems of other industries, including brass-founding in Bristol, was Abraham Darby. In 1707 he patented a method for casting iron pots in sand, then in the next year moved from Bristol to Shropshire and leased a blast furnace at Coalbrookdale, where he was soon making iron pots on a considerable scale. Almost immediately, it seems, he started experiments with first coal and then coke in the furnace, and probably had developed a successful process using coke by 1709. However, coke blast furnaces were not adopted by other ironmasters for a very long time. In 1750, there were only three such furnaces in England, compared with seventy charcoal furnaces, and all three belonged to the Darby family. One reason was that although the quality of iron produced in the coke blast furnaces was ideal for making cooking pots, it was not so good for conversion to wrought iron, the purpose for which most charcoal iron was made. The latter also remained competitive because the cost of charcoal production in existing coppiced woodland appears to have been fairly stable. Only after the middle of the century did coke blast furnaces in coal-mining areas achieve markedly lower costs.

Not only was Shropshire the first iron-producing region in Britain (and indeed, Europe) to use coke in blast furnaces, but the Shropshire industry was also a pioneer in making components for steam engines and then in applying the engines to iron industry processes.

Newcomen's first engines had brass cylinders, but the techniques that contributed to Abraham Darby's success in making large, thin-walled cast-iron pots meant that he was ideally placed to make engine cylinders in cast iron. The first ones produced in Coalbrookdale were cast in 1718, then from 1722 cylinders were regularly made there.

At this point, one needs to recall that in order to maintain the continuous blast of air on which the operation of a blast furnace depends, waterwheels were employed to work large bellows. Water stored behind a dam above the Coalbrookdale furnaces kept the bellows working, but in dry summer weather, there was often insufficient water to operate the furnaces. To overcome this problem, a Newcomen engine was installed in 1742 to pump water that had already been used by the waterwheels and return it to the reservoir above the dam. Soon, however, thought was given to designing a steam-driven air pump that could replace the water-powered bellows. This would allow a steam engine to provide the air blast for a furnace directly. This was probably first achieved at a Shropshire ironworks owned by John Wilkinson in 1776.

Developments in the iron industry were now closely linked to advances in steam engine design, especially at Wilkinson's works. Most strikingly, the success of James Watt's improved steam engine with its separate condenser depended on obtaining a cylinder with a uniform bore in which the piston could move with minimal leakage of steam around its edges. Cylinders made at Coalbrookdale were unable to meet this standard. However, the Wilkinson ironworks was manufacturing heavy guns and possessed a machine, patented in 1774, for boring out gun barrels. It proved also to be capable of boring out engine cylinders with the accuracy Watt required, and Wilkinson thus became a major manufacturer of engine parts. He also pioneered the application of Watt engines for working forge hammers (1782) and for driving rolling mills (around 1790).

The most famous achievement of the Shropshire ironmasters was to build a high arched bridge across the river in the dramatic, rocky landscape of the Severn Gorge. This was not the first iron bridge in the world. There were much earlier ones in China, and two iron footbridges had been built in northern England not long before.[2] However, nearly all these earlier bridges were made of wrought iron, and many were suspension bridges with iron chains. The iron bridge over the Severn, whose components were made at Coalbrookdale between 1777 and 1780, was the first large-scale demonstration in Britain of the potential of cast iron. The idea for such a bridge had been discussed by an architect named Pritchard and was supported by John Wilkinson, but it was the Coalbrookdale ironworks owned by the Darby family that made and erected it.

The bridge was immediately famous. It was mentioned in advertisements for coaches that passed nearby. Pictures of it were widely sold. Its image decorated locally made pottery, and in 1782 one enthusiast described it as like an "elegant Arch in some ancient Cathedral." Noting such comments, an important recent study argues that the bridge became a major "symbol" of industrial development, representing the potential of technology, and suggesting an "optimistic" vision of the future of industrial society.[3]

Throughout these episodes, interdependence of innovations in ironmaking, the use of coal, and steam power is striking. In its earliest form, Newcomen's engine was of such low thermal efficiency (0.5 percent) and consumed so much fuel that it could hardly have been a success anywhere but at a coal mine, where fuel was very cheap. But at the same time, many coal mines were in urgent need of better means of pumping water, which the engine provided. Similarly, the iron industry needed to use steam engines, but also contributed to the development of cast-iron engine cylinders.

These close relationships between industries and between innovations invite comparison with patterns of innovation characterized as "movements" in technology in previous chapters. In eighteenth century Britain, coal, steam power, and the coke-based iron industry were so closely related that it is useful to think of them as a movement in this sense. It was not the only significant movement of the period, however. Two others were discussed in the previous chapter, one concerned with agricultural improvement and the other with making rivers navigable, building canals, and constructing other public works. Yet another movement, ultimately of much greater significance, was related to the textile industries, which comprised well-established wool and linen manufactures and relatively new enterprises in silk and cotton. Initially, the silk industry was a notable source of new ideas but within a century, every branch of textile manufacture had been affected.

Textile Technology and Organization

Most textile industry at this time was organized on the domestic system, with individual spinners, weavers, and stocking knitters working in their own homes using manually driven equipment: spinning wheels, hand looms, and since about 1650 in the Nottingham and Leicester area, the stocking-knitting machine mentioned in chapter 5.

However, a radically new concept was introduced after 1700 when the water-powered silk mill mentioned earlier was built at Derby. Its function was to manufacture silk yarn from raw silk, the yarn being used by local stocking knitters to make silk stockings. At first, Dutch silk-twisting machines were used in the mill with only moderate success. Then John and Thomas Lombe acquired the enterprise and set out to discover more about the long-established silk mills of northern Italy, an early form of whose machinery was illustrated in chapter 2 (figure 7). When the Lombes reorganized the Derby mill between 1717 and 1721, they equipped it with copies of the Italian silk-twisting or "throwing" machines. The mill was now so successful that other factories of the same kind were established, one other at Derby being part-owned from 1762 by a certain Jedediah Strutt.

Demand for cotton textiles was expanding more rapidly than demand for the more expensive silk, however. Efforts to increase output of cottons took two directions. One was to speed up production *within the domestic system.* Devices such as Kay's flying shuttle (invented in 1733 but not much used until the 1760s) allowed hand-loom weavers to work much faster, while the spinning jenny (invented about 1764 by James Hargreaves) greatly increased the productivity of home-based spinners.

The second direction taken by innovation in cotton manufactures was prompted by the Derby silk mill, which raised the question as to whether similar methods could be used for the spinning of cotton thread. For example, several features from the machines used in the silk mill were adopted by Paul and Wyatt in a cotton-spinning machine they devised in 1738. Then comparable ideas were developed further in the 1760s by Thomas Highs of Leigh in Lancashire and his associate John Kay, not the inventer of the flying shuttle but a clockmaker.

Richard Arkwright was also a Lancashire man. He had worked as a barber and wigmaker at Preston and then at Bolton. Little else is known of his early career, but he evidently heard of the work of Highs and Kay and enlisted Kay's help in the construction of a similar machine. This was successfully completed by 1769, when Arkwright obtained a patent for a spinning machine.

At this stage, Arkwright set up his first factory in which cotton spinning was carried on by means of power-driven machinery. This was at Nottingham, where the hosiery industry promised a good market for Arkwright's yarn. Power was provided by horses. Probably six were used at a time; they were harnessed to a horizontal wheel, 27 feet (8 meters) in diameter, which was housed in a ground-floor room of the factory. Wooden shafts and pulleys took the drive from this wheel to the spinning machines on the floors above.

Arkwright raised the capital needed to start this factory by entering into a partnership with two local stocking manufacturers, one of them being Jedediah Strutt, whose silk mill has been mentioned. The firm prospered and in 1771 a larger factory, powered by waterwheels, was set up at Cromford, two miles south of Matlock in Derbyshire. Within a short time 600 workers were employed there. Then, in the late 1770s, Strutt and Arkwright collaborated to set up new mills at Belper and Milford, a few miles north of Derby, both with waterwheels driven by the River Derwent. By means of other partnerships Arkwright became associated with four more cotton factories in Derbyshire, and others in Lancashire and Scotland. All these were in operation by 1790, and their construction so soon after Arkwright's first invention gives a measure of how quickly the cotton industry grew.

The importance of Arkwright's work is not only that he mechanized all the processes involved in producing cotton yarn, from carding the raw cotton to spinning the finished thread. Beyond that he saw that the whole system of production would be better controlled if it were centralized in a single factory rather than being dispersed among the homes of large numbers of spinners. In other words, he changed the organization of the industry from the "domestic system" to the "factory system." Again, there is evidence that the Derby silk mill had established a precedent, this time regarding aspects of factory organization such as shift work and discipline.[4]

The movement in technology and industry represented by the new textile factory system had so far no connection at all with the developments in steam power and ironmaking discussed earlier. It originated in a different region of England, centered on Nottingham and the eastern fringes of Derbyshire, where leading men in the stocking-knitting industry seem to have been especially interested in the potential of new techniques. Both Hargreaves and Arkwright had originated in the Lancashire textile area but moved to Nottingham to develop their ideas. As we have seen, the success of these developments in Nottingham and Derbyshire soon led to emulation in Lancashire and elsewhere, but these were still very much regional industries, little influenced by developments in other regions such as Shropshire, the seat of the cast-iron industry.

The distinctiveness of this "movement" in textile technology is further illustrated by the fact that the machinery in the silk mill and in Arkwright's early cotton mills was largely built of wood and was driven by waterwheels (or in a few cases by horse gins). The factory buildings and the waterwheels themselves were of traditional construction and were merely enlarged versions of the structures that country millwrights had been building for centuries. Thus the new textile industry, however novel in mechanical detail and overall organization, was still using the old preindustrial resource base—timber, horses, water power—in complete isolation from the movement described earlier that was developing a new resource base—iron, steam power, coal.

The Convergence of Industrial Movements

As the Arkwright system of cotton spinning continued to expand, it became more and more difficult to find convenient sites for factories near to rivers capable of driving waterwheels. In Manchester, the commercial center of the Lancashire cotton industry, a steam engine was installed in 1783 by one of Arkwright's partners at a cotton mill. It worked a pump and so augmented the water supply available to power the mill. Matthew Boulton, James Watt's partner, reported that Manchester men were becoming frantic for a more satisfactory way of using steam power to drive their machinery. They were "steam mill mad," he said, urging Watt to devise a steam engine adapted for this purpose.[5]

At the same time, the inadequacy of traditional building structures was becoming evident as wooden floors sagged under the increasing weight of machines and as fires destroyed factories disturbingly often. More and more iron was then used in factory buildings, first in the 1780s with the introduction of cast-iron columns to give extra support to timber floor beams, and then from 1792 in new methods of construction designed to make buildings fireproof (figure 29). By 1805 iron was replacing timber in the construction of many types of factory machinery, including waterwheels and the beams and framing of the steam engines.

The combined use of steam power and iron construction enabled the factory-based textile industry to break through the limits that reliance on traditional resources would otherwise have placed on it, and the expansion of textile manufactures in turn stimulated the iron and engine-building industries by creating demands for new products. There was thus a convergence between two originally quite distinct industrial and technological movements that contributed to launching both of them into a phase of accelerating expansion. It is this convergence and the consequent growth of new modes of production that more than anything else justifies the term "industrial revolution."

Statistics that illustrate the transition from evolutionary to revolutionary phases in these two industries are plotted in figures 30 and 31. The growth of the cotton industry is approximately documented by the quantities of raw cotton it consumed, all of which had to be imported. It will be seen from figure 30 that imports began to increase more steeply from about 1775, soon after Arkwright's mills came into full production—and after a duty that had restricted sales of cotton goods in Britain had been partly lifted by Parliament. A rapid development of export markets also contributed to the growth in cotton manufactures.

Acceleration in demand for iron goods lagged about thirteen years behind and began to take off in the middle 1780s just as the builders of new textile factories were increasingly installing

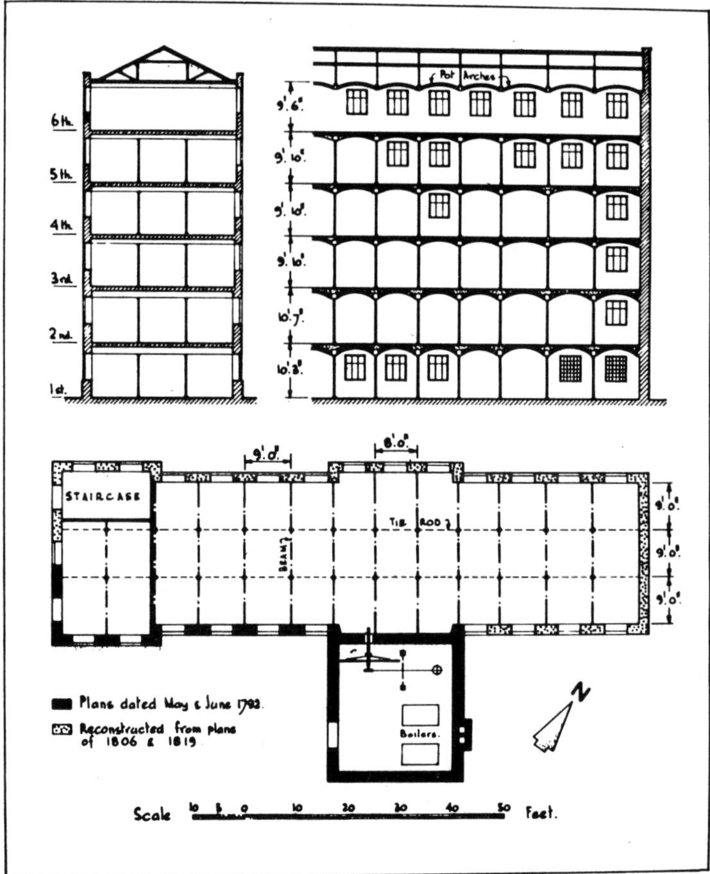

Figure 29
William Strutt's fireproof structure. This six-story cotton mill designed by Strutt was erected at Derby in 1792–93, and was the first multistory fireproof building. The floors consisted of brick arches carried on heavy timber beams, which themselves were supported by two rows of cast-iron columns.

(Source: H. R. Johnson and A. W. Skempton, "William Strutt's cotton mills," in *Newcomen Society Transactions, 30,* 1956. By courtesy of the Newcomen Society).

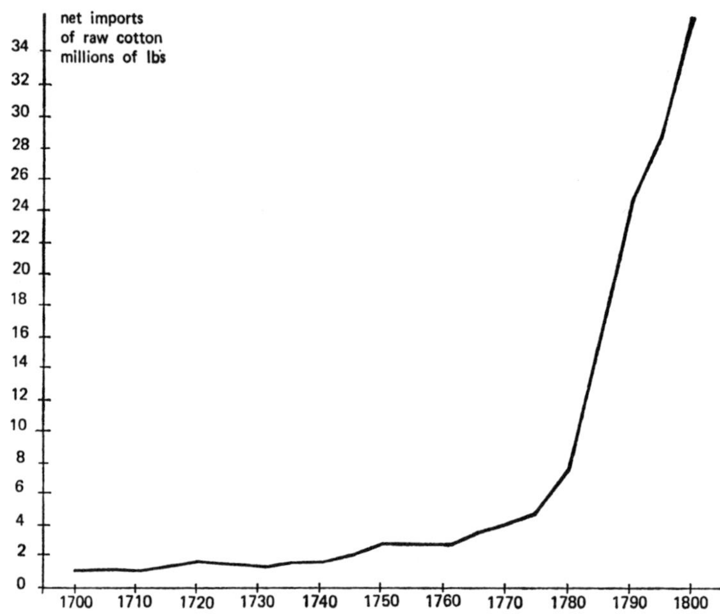

Figure 30
Raw cotton used in the British textile industry during the eighteenth century. Figures are in pounds (lbs) and have been calculated to give ten-year moving point averages.

(Sources of data: A. P. Wadsworth and Julia de Lacy Mann, *The Cotton Trade in Industrial Lancashire,* Manchester University Press, 1931, pp. 520–521 (appendix G); also M. M. Edwards, *The Growth of the British Cotton Trade, 1780–1815,* Manchester University Press, 1967, p. 250; and Phyllis Deane and W. A. Cole, *British Economic Growth 1688–1959,* Cambridge University Press, p. 51. Graph drawn by the author.)

steam engines and iron columns. Demand from the textile industry does not account for all the increase, however. More significant was an enormous demand for cast-iron cannon and wrought iron for small arms manufacture during the Napoleonic Wars. At the same time, more iron was being used to make colliery railroads, and from 1795 several more iron bridges were built.

Britain's rapid industrial development had impacts in many other parts of the world. For example, cotton was hardly known in Britain until imported cloth began to arrive from India in the seventeenth century. This proved extremely popular, partly because it was so much finer and lighter than most traditional wool and linen textiles and partly also because of the bright

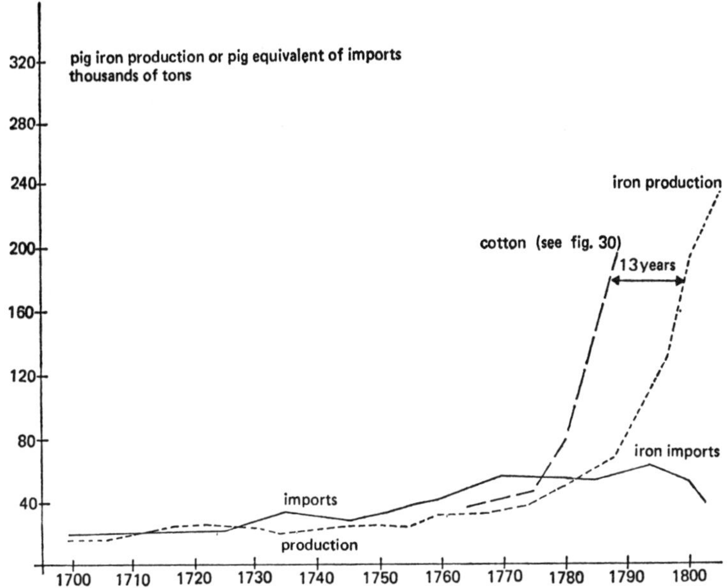

Figure 31
Iron production in England and Wales, together with imports, during the eighteenth century.

Dashed lines represent the author's estimates, based on analysis of all available figures for imports, exports, and consumption (e.g., given by Phyllis Deane and W. A. Cole, *British Economic Growth,* Cambridge University Press, 1962; B. R. Mitchell, *Abstract of British Historical Statistics,* Cambridge University Press, 1962; C. K. Hyde, *Technological Change in the British Iron Industry,* Princeton University Press, 1977).

colors with which much of it was dyed—colors that remained remarkably "fast" when the cloth was washed. The high quality of this cloth presented a challenge of considerable technological significance to British spinners, weavers, and dyers, which for a long time they could not meet. But with the growth of production in Arkwright's factories and the invention in 1785 of the "mule," a new spinning machine that produced extremely fine yarn, India began to lose markets in Britain and in the countries in which the British exported cloth, ultimately with very damaging results.[6]

Another more positive impact of the new industry was felt in the Caribbean and American plantations from which much of the raw cotton used in Britain was beginning to come. Bales of cotton exported to Britain from the southern United States rose from less than 300 in 1792 to a staggering 36,000 in 1800. This rapid development was assisted by the invention of a new cotton gin around which a good deal of legend has accumulated.

Cotton textiles had, of course, originated in Asia, and it is not surprising that the first cotton gins were invented there. Indeed, a simple cotton gin was introduced into Louisiana from western Asia in 1725 and worked well for long-staple cotton. What was invented in America about 1792 and patented by Eli Whitney was a gin that could separate the seeds from the fibers of short-staple cotton. Like the traditional Asian machine, it consisted of rotating cylinders, but now hooks on the cylinders passed through slots that held back the seed.

The Technology of the Industrial Convergence

When Matthew Boulton wrote to James Watt reporting that Manchester cotton-mill owners were becoming desperate for an effective steam mill, the two men had been in partnership for about seven years. Watt had patented his first improvement of the steam engine in 1769. This was the separate condenser that eliminated the waste of heat that occurred in the Newcomen engine when steam was condensed inside the cylinder. But there were several difficulties in turning this idea into a practical engine, and before the partnership with Boulton was agreed in 1774, Watt's lack of capital had limited his ability to overcome them.

It has to be stressed that during the 1770s, Watt's improved engine was still purely a pumping engine and met with particular success in the tin- and copper-mining areas of Cornwall, where coal was expensive and the fuel economy of the engine was especially attractive. In order to adapt the engine to produce rotary motion, and so to drive cotton-spinning machinery, Watt chose to make it "double-acting." The point here is that pumping engines generated power only when the piston was being pushed downward, whereas for reasonably smooth rotative motion it was desirable for the engine to produce power during both the upward and downward strokes.

At this point, Watt was faced with two problems concerned with the motion of the engine's parts—that is, problems of "kinematics." The engine retained a pivoted, overhead beam, like the beam of the Newcomen engine and all contemporary pumping engines. This meant that the motion of the piston had first to be transformed into the swinging motion of the beam. Then the swinging beam had to turn the flywheel (figure 32).

The connection between the beam and the flywheel could most obviously be made by means of a rigid connecting rod and a crank. However, Watt thought that his use of a crank would be restricted by a patent that had been taken out by somebody else. Wishing to avoid trouble over this, he considered four other methods for linking a connecting rod to a flywheel without using a crank. In 1781 he patented all four of these methods, although only one of them, the "sun and planet" motion, was used regularly on his engines (figure 33).

The other kinematic problem was how to transform the straight-line reciprocating action of the piston into the swinging motion of the beam. This was more difficult than in the traditional pumping engine. The chain and arch-head motion used on the latter (and clearly illustrated in figure 23) could not be used with a double-acting engine because the chain would sim-

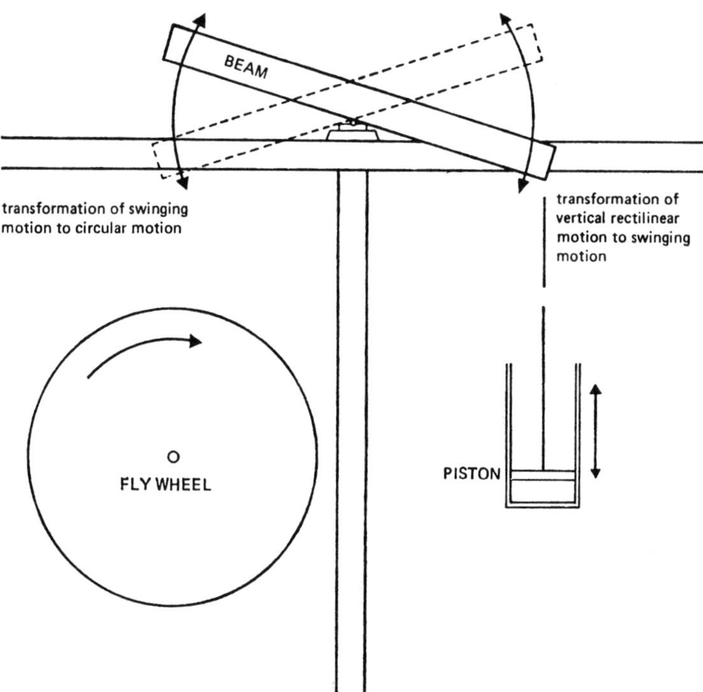

Figure 32
The kinematic problems in Watt's rotative engine. (Diagram by D. M. Farrar.)

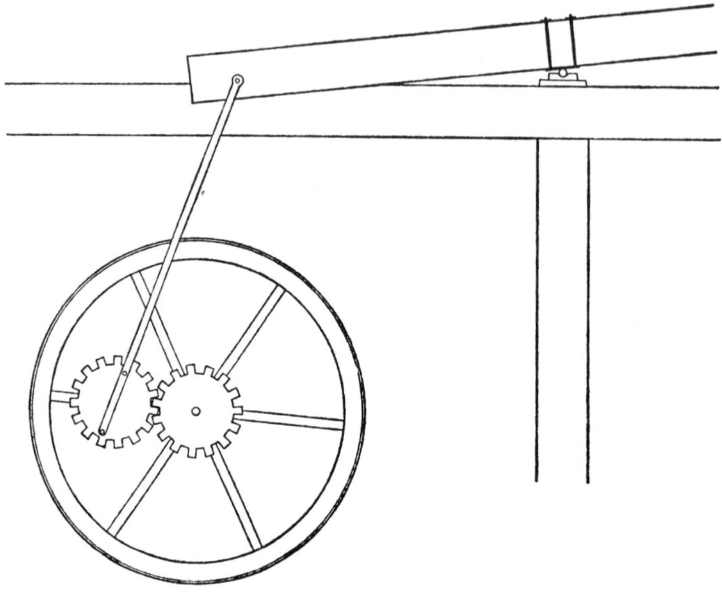

Figure 33
Solutions to kinematic problems (1): transformation of a swinging motion into a rotary motion by the "sun-and-planet motion" used on Watt's engines. (Diagram by D. M. Farrar.)

ply go slack without transmitting power when the piston was pushing upward. Yet the piston rod had to move vertically through the stuffing box set in the cover that was now necessary to close the top of the cylinder. It could not be allowed to swing from left to right in following the motion of the beam. The answer was yet another kinematic invention, Watt's "parallel motion." However, it is an interesting reflection on Watt's mode of thinking that he again considered alternative methods of achieving the same result (figure 34)—and patented them to block the path of possible rival inventors.

Watt's rotative steam engine was thus the result of several inventions—the separate condenser, patented in 1769, and then double-acting operation and the several kinematic inventions, patented in 1782 and 1784. The first engines of this kind to be used in the textile industry were installed at cotton mills near Nottingham (1785) and in Manchester (1789), raising problems about measurement of power that encouraged Watt to begin specifying the output of his engines in "horse-power."[7]

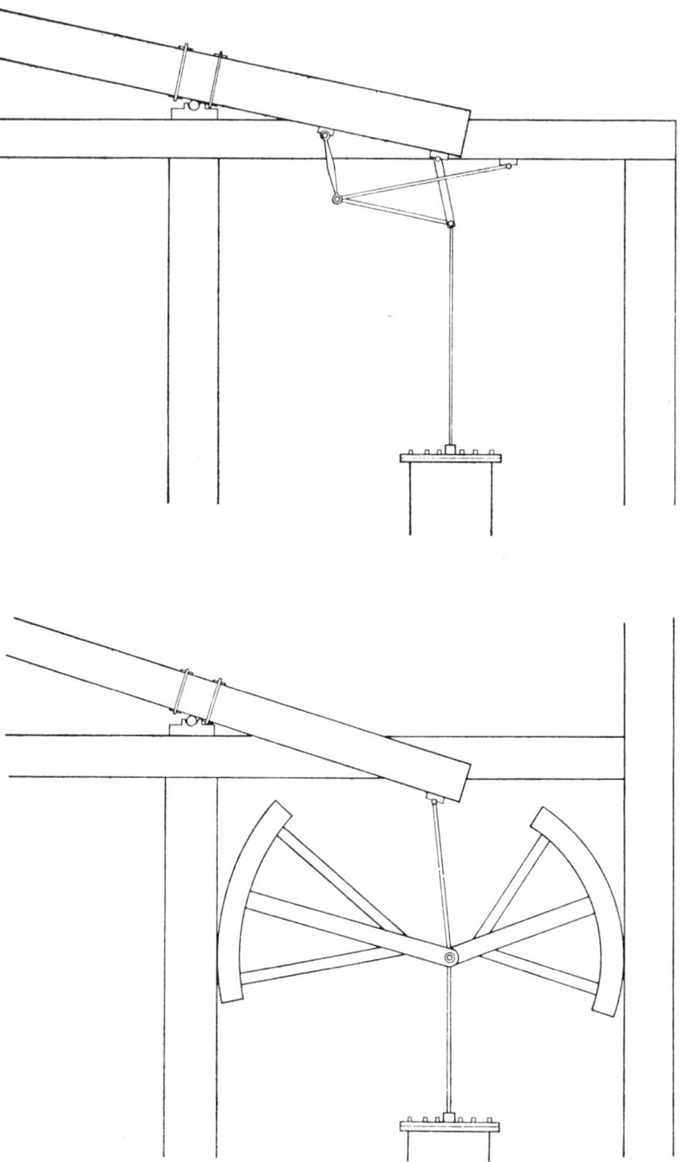

Figure 34
Solutions to kinematic problems (2): transformation of a vertical reciprocating motion into a swinging motion. The fact that Watt was able to devise several solutions, of which two are shown here, implies an ability to organize his thinking about kinematics in a rational manner rather than relying solely on an intuitive approach. The top drawing shows the "parallel motion" actually used on Watt's engines (compare figure 35). (Diagram by D. M. Farrar.)

What was more remarkable than anything else, however, was the impact of the rotative engine on Manchester. Industrial development had previously been limited there by the town's rather flat site, which offered few opportunities for building water-powered mills. With the arrival of the rotative steam engine, many more textile mills could be built, taking advantage of coal supplied by canal boat from nearby mines. Boulton and Watt installed several engines in the 1790s, but local firms also began to build them.[8] Thus Manchester was transformed in a few years from a small but important commercial town to the biggest concentration of manufacturing and engineering industry in the world, with important engine-building and machine-making establishments as well as numerous cotton mills. Here, more than anywhere, the steam, coal, and iron "movement" in technology that began with Darby's coke-blast process and the Newcomen steam engine in 1709 and 1712 converged with the textile factory "movement" that we can date from the relaunch of the Derby silk mill between 1717 and 1721 and Arkwright's patent of 1769.

Another aspect of this convergence of two movements in technology was the wider use of iron in the construction of machines and buildings. One of the pioneers was William Strutt, a son of Arkwright's former partner Jedediah Strutt. Concerned about devastating fires in some cotton-spinning factories, he devised the fireproof structure mentioned earlier in which the upstairs floors in a factory were made as a series of brick arches. This can be seen in the transverse section of one of Strutt's mills, illustrated in figure 29 at top right, which also shows how the heavy floor structure, still incorporating timber beams, needed numerous cast-iron columns for its support. Strutt's friend, Charles Bage, adopted the same principle in a flax-spinning mill he built at Shrewsbury in 1796, but now with the refinement of cast-iron beams to support the brick arches. After 1800, Strutt and Bage erected further factory buildings with iron beams and columns and fireproof floors, making several further improvements.[9]

Cast iron was a novel structured material, and engineers felt some uncertainty about how to design for its use. In 1795, Thomas Telford proposed to build a cast-iron bridge and also an iron aqueduct in Shropshire. His designs for both structures seem to have involved much discussion with William Reynolds, partner of the Darby family at Coalbrookdale. To test the strengths of the proposed spans, Reynolds had models of key components cast in iron. These were supported in positions comparable to the way they would be used in the completed bridge, and weights were placed on them to represent the loads to which they would be subjected. A calculation was then possible to determine what the corresponding breaking load on the full-size bridge would be.[10]

Charles Bage, designer of the Shrewsbury textile mill just mentioned, was in touch with Telford and used the results of Reynolds's tests in designing columns and beams for the mill. One example was a test on the breaking of iron bars, 3 feet long and an inch square (that is, 915mm long and 25mm square) that provided data from which Bage could estimate the strength of his much larger beams. He also carried out tests of his own.[11]

What is of very considerable interest, however, is that in calculating how the strength of the full-scale beam or bridge would relate to the test on a model, Bage[12] and Reynolds used a theory derived from Galileo's work on the strengths of beams. It will be recalled from chapter 3 that Galileo had formulated in mathematical terms how the strength of a model would compare with the strength of a structure of larger size, made from the same material. This analysis was well known in England from a book by William Emerson[13] whose fourth edition of 1794 was probably consulted by Reynolds, Bage, and Telford. Further evidence of the widespread use of the Galileo theory is provided by another volume, published in 1806, in which Olinthus Gregory acknowledged the existence of more advanced theories in the work of continental mathematicians such as Euler. These were "certainly very ingenious; but we prefer the comparatively simple theory of Galileo."[14] Shortsighted though this attitude was, it could be justified for so long as theory was needed only for comparisons between models and large structures. Only with the advent of the long-distance railroad in the 1830s did a more sophisticated approach to the design of iron beams, girders, and bridges become necessary.

As iron columns and beams were more often used in factory buildings, so experiments began in using cast iron to make the overhead, pivoted beams for steam engines. The pioneers were associated with the little-known Yorkshire firm of Aydon and Elwell, in 1795–96, and seem to have obtained advice on theoretical aspects from John Banks (c. 1740–1805), an itinerant lecturer on mechanics and "natural philosophy." It is clear that

Banks, like Bage, made use of Galileo's theory, emphasizing not only the aspects that referred to models, but also Galileo's idea that a beam of optimum economy and strength would have a parabolic profile. This idea could be directly applied to an engine beam. If the beam was rectangular in shape it would be highly stressed at its midpoint, but only lightly stressed toward the ends. Moreover, in working with a heavy and expensive material such as cast iron, it made sense to take note of the implications of this and design the beam to be deeper in the middle than at the ends. As in Galileo's example, the ideal profile would be parabolic, but when two parabolas were combined, a reasonable interpretation of the optimum shape was an ellipse. Banks experimented with various shapes for engine beams at the Aydon and Elwell foundry, and testing a series of models, confirmed that the elliptical shape was strongest in proportion to its weight.[15]

Soon after this, Boulton and Watt engines began to be built with elliptical cast-iron engine beams. The firm may have been prompted by one of Watt's Manchester friends, George Lee, partner in a cotton-spinning firm that was quick to adopt such innovations. Thereafter, for as long as beam engines were built, their beams were almost invariably elliptical, in striking tribute to the practical relevance to Galileo's mechanics (figure 35). Similarly, cast-iron floor beams in factories commonly had a gently curved, humpback profile, deepest at midpoint and approximating to the parabolic shape.

Apart from his work on engine beams, John Banks also conducted experiments in which he measured the varying steam pressure within a working engine's cylinder. In one of his books he illustrated the pressure gauge he used, but it is not clear that he obtained results of any great significance. Somewhat earlier, Watt's friend George Lee employed similar gauges and also several thermometers in studying the performance of a Boulton and Watt engine at his cotton mill near Manchester. The experiments extended over two years (1793–95), and Lee investigated a wide range of topics, some related to fuel economy and the power needed to run particular types of machine, and some related to scientific concepts and current views about how heat "flowed" through a working engine.[16] Although Lee was playing with purely scientific ideas here, his experiments were mostly directed toward a practical problem that had worried Watt ever since he began to design rotative steam engines for textile mills.

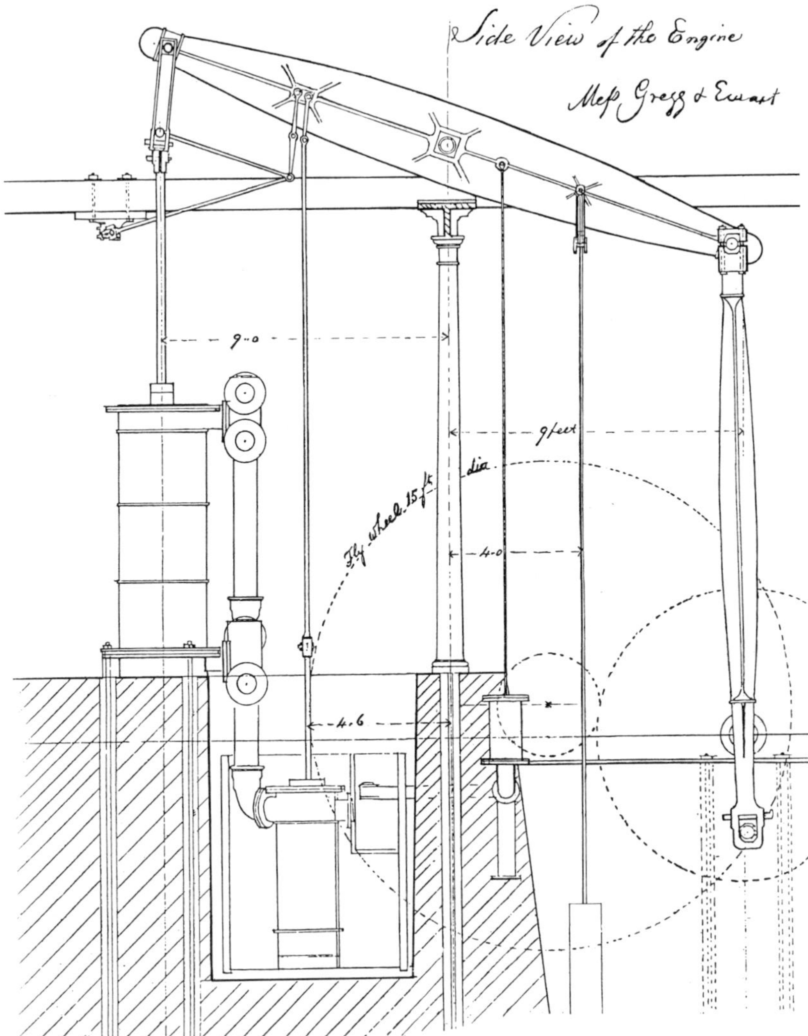

Figure 35
A double-acting rotative steam engine built by Boulton and Watt in 1802 at a cotton-spinning mill near Manchester owned by Messrs. Greg and Ewart.

This engine has a cast-iron beam, pivoted at the center, each half of which conforms rougly with the "parabolic profile" recommended by Galileo. The overall effect is an approximate ellipse. The left-hand end of the beam is connected to the piston rod via Watt's parallel motion. The engine's separate condenser, located in a pit to the right of and below the cylinder, has a connecting rod from the beam working its associated air pump. In this engine, the flywheel was driven through gears (shown in outline only) via a crank and not by the sun-and-planet motion. The connecting rod linking the right-hand end of the beam to the crank is of a flattened elliptical shape, again showing the influence of the "parabolic profile" concept.

(From a drawing of 1802 in the Boulton and Watt papers. Reproduced by permission of the Reference Library Archives Department, Birmingham Public Libraries.)

There was no easy way to estimate how much power would be needed to drive a particular group of machines, and hence it was difficult to decide how big an engine was needed by a particular factory.[17]

Historians have taken widely different views about whether "science" in any accepted sense contributed to the technology of the industrial revolution.[18] It can easily seem that many key inventions were made by craftsmen who knew nothing about it. However, James Watt, George Lee, William Reynolds and even John Banks had a good knowledge of science. Moreover, some of the technology affected by the convergence of iron and steam power with the textile industry was presenting problems of a new kind for which no craftsman had relevant experience, such as establishing adequate dimensions for iron beams or mill engines. It was almost inevitable, then, that the people who met with most success in these new circumstances were those who could use some relevant scientific ideas to compensate for unavoidable lack of experience.

Rationalist Ideals

While for some people, the practical use of scientific ideas may have been an entirely pragmatic choice, for others, there was an element of idealism about it. This makes it more difficult to answer questions about whether or when science was effectively applied, because many people attended scientific lectures or did simple experiments in their homes because of an ideological belief that knowledge of nature and rational inquiry was important, even if the experiments they did were actually quite irrelevant to their work.

At the end of the eighteenth century, one of the Darby family carried out experiments with a camera obscura and an electrical machine, neither of which were of industrial relevance.[19] In Birmingham, Watt and Boulton met with other men interested in science at meetings of the Lunar Society, while in Derby and Manchester, more formally constituted and longer-lasting "literary and philosophical" societies were supported, around 1800, by such men as William Strutt and George Lee, two cotton manufacturers mentioned above. Several of these men were using some knowledge of chemistry or mechanics in their work, but much of what they discussed in these societies was probably on the level of science as a cultural interest.

In the previous generation, however, the Yorkshire engineer John Smeaton not only believed in the relevance of a scientific approach but consistently brought it to bear in the way he practiced engineering. He had begun life as an instrument-maker and astronomer, but his classic work was a long series of experiments on the performance of waterwheels, conducted with the aid of a carefully made model. His account of the experiments was published in 1759 and reprinted several times.[20] There is ample evidence that it was widely read, not least by several of Watt's friends. It demonstrated the superior efficiency of the so-called breast-shot waterwheel as compared with the traditional undershot wheel, and so helped establish the type of waterwheel that beame standard in early textile mills.

Smeaton showed the quality of his approach in a discussion of his experimental results where he quoted "laws of reasoning by induction."[21] It is clear from the context that these words refer to Isaac Newton's "rules of reasoning" about "propositions inferred by general induction." Newton had discussed these rules in his great book *Principia* (1687) as a preliminary to comparing a large number of astronomical observations with calculations based on his theories of planetary motion.[22] Smeaton, as an astronomer, was familiar with Newton's book. He possessed a copy (in Motte's translation), and his writings show knowledge of it in several places. In the work on waterwheels, he clearly saw that discrepancies in his experimental measurements involved the same methodological problem as Newton's comparison of theory and observation in astronomy.

Throughout Smeaton's career, he always stressed the importance of careful investigation, measurement, and testing before engineering projects were implemented, and he trained engineers who worked with him to adopt the same approach. Smeaton was aware that in France engineers were given formal training in mathematics, and he was very concerned that in Britain there was "no public establishment, except common schools, for the rudimental knowledge necessary to all arts, naval, military, mechanical and others." The education of engineers was instead "left to chance."[23] In order to give some shape to the emerging engineering profession and perhaps remedy some of these problems, he encouraged the formation of a "Society of Civil Engineers" in 1771, and it is noteworthy that its members informally referred to themselves as "Smeatonians." They adopted a motto that expressed the ideal of rational methods and measurement in engineering. It consisted of words from an apocryphal book of the Bible where God is described as having ordered "all things in measure, number and weight." In like manner should engineers order their work.[24]

In attempting to show how rational methods and calculation could be applied to engineering, Smeaton contributed greatly to the emergence of engineering as a *profession,* in contrast to the old craft-based methods of former generations of mill-wrights and bridge-builders. But historians who argue that science had little relevance to industrial technology in Britain at this time can rightly point out that Arkwright's spinning machine and many other of the new devices seem to have depended for their realization solely on the mechanical ingenuity of a succession of fairly ordinary craftsmen. It would appear that most of these people had no special scientific or theoretical knowledge, but were simply "mechanically minded."

John Smeaton never had to deal with the new textile machinery, nor was he connected with any of the notable mechanical inventions of the period. Therefore he never had the stimulus to apply his rationalizing impulse to the processes of mechanical invention. In the course of time, however, the ideas involved in the invention and design of machines did come to be rationalized and codified, and textbooks were written about what became known as the "kinematics of mechanism."[25]

The need to formalize the subject grew more urgent when this aspect of engineering had to be taught to students. The origins of "kinematics" as a science must therefore be sought partly in early attempts to provide organized technical education. In 1794–95, the *École Polytechnique* was founded in Paris as part of a scheme for enlarging and coordinating the work of the existing technical schools. It is clear that some form of machine kinematics was taught at this new institution as part of a course on descriptive geometry.[26] The syllabus was devised by Gaspard Monge (1746–1818), the former teacher at the *École du Génie* who is celebrated for his innovations in the field of technical drawing. He classified the motions of machines using two pairs of basic categories: circular or rectilinear motion, and continuous or reciprocating motion.[27] A little later, two other authors writing in Paris had much the same idea: "The motions which are used in the arts are either rectilinear, or circular, or determined by some given curve; they can be either continuous or reciprocating. . . . All machines have as their object the transformation or communication of these motions."[28] At the end of their book, these authors gave a diagram illustrating their classification of the motions of machines. This included some mechanisms that were clearly derived from a knowledge of Watt's engines.

It may seem out of place to mention these French authors here, but it was a feature of the industrial revolution that Frenchmen wrote the best books about the new technologies that had been pioneered in Britain. Watt and Smeaton had both found it necessary to study French books on engineering because there were so few English works on the subject. Even in the next generation it was French and German authors who succeeded best in formulating a rationalized approach to engineering.

Social Ideals and Industrial Technology

One of the most perceptive historians of the industrial revolution, D. S. Landes, sees two major impulses behind the innovations of this period. One is "the application of rationality to life," some aspects of which we have seen in previous pages. The other is what he calls a "Faustian ethic," which is the urge to assert mastery over "nature and things."[29]

In many paintings by eighteenth-century artists that depict the pioneer iron bridge near Coalbrookdale in Shropshire, rationality is expressed by the geometrical lines of the bridge itself, but aspects of nature that remain unmastered are indicated by emphasis on the wildness of the bridge's natural setting. Contemporary paintings of the Coalbrookdale blast furnaces often showed them as they appeared at night, emphasizing the elemental force of fire mastered by man.[30]

The question remains, though, for what social purpose was nature being mastered? The previous chapter quoted poets of the early and middle eighteenth century who described the control of elemental forces by works of engineering, but stressed that this was a means of improving public amenities (as the "broad Arch" contained the dangerous flood), and promoting the public good (as engines pumped water to "thirsty cities").[31]

Public amenity continued to be stressed in the early phases of the industrial revolution as canals and turnpike roads were built. These were works of the kind on which the engineer John Smeaton often worked, and Smeaton is said to have been the first person in Britain to call himself a "civil engineer." The word "engineer" currently had prestige only when applied to military engineering, and he wished to distance himself from that. As his daughter, Mary Dixon, commented, he felt he owed "a debt to the common stock of public happiness or accommodation."[32]

We may compare the latter comment with the interpretation of the Coalbrookdale iron industry offered by Abiah Darby, daughter-in-law of the pioneer of the coke-fired blast furnace. She saw the innovations made by members of her family as "achievements of benefactors of mankind, to be contrasted with those of soldiers, the destroyers of mankind."[33] As with Smeaton, public benefit was stressed by distancing the technology concerned from military applications. The Darby family were Quakers, and from the outbreak of the Seven Years' War in 1756 they refused all orders for cannon in conformity with the Friends' peace testimony, even though neighboring ironworks were prospering mightily from such orders.[34]

In the two previous chapters, a contrast was drawn between values such as these, which stressed the public good, and the ethic that justified the use of economic self-interest as a guide to the best use of wealth. In the second half of the eighteenth century, conflict between these viewpoints may have seemed to be resolved. The pursuit of rationality in engineering, industry, and society was serving both the public interest and the self-interest of entrepreneurs. It was leading to so much constructive development and promised so much in terms of improved public amenity, social order, and general prosperity that it was hard to feel anything but optimism.

On a more philosophical level, this optimism seemed to be confirmed by the former professor of moral philosophy at Glasgow University who is famed for his *Inquiry into the Nature and Causes of the Wealth of Nations,* published in 1776. In this and earlier writings, Adam Smith argued that it was futile simply to deplore the predominance of self-interested behavior in trade and commerce, for such behavior was a natural part of life and often had a constructive effect. Benevolence might be the greatest virtue, but self-interest could lead to lesser virtues such as thrift, hard work, and discretion—virtues central to the so-called Protestant ethic.

Taking a realistic view, then, it is not "from the benevolence of the butcher, the brewer, or the baker that we expect our dinner, but from their regard of their own interest."[35] There was no disadvantage in this, because there was no essential conflict between the butcher's self-interest and that of his customer. The latter was satisfied if he got a good piece of meat at a reasonable price, while the butcher attracted more customers and made bigger profits by consistently supplying good meat. The natural order of things in society ensured that in a free market everybody got the fairest possible treatment. The problem was that governments could prevent the "natural" law of the free market from working for the common good by introducing arbitrary rules and regulations.

When Adam Smith visited France during 1764–65, he found similar arguments being used by a group of people who used the term "laissez-faire" to describe the policy of allowing economic life to develop freely without government interference. These men were sharp critics of the mercantilist policies then being pursued by France (which were described in chapter six), and they wished to see a more "natural" form of development in agriculture and trade, less hampered by cumbersome regulations.

Toward the end of the eighteenth century, arguments of this sort were put forward with real conviction. They were not just cynical attempts to whitewash the selfish actions of factory owners and traders, or to protect them from government intervention. They were, rather, the product of a genuine belief in a "natural" order in society, accompanied by a remarkable and optimistic expectation that if "nature" was allowed to take its course, there would be steady social improvement. Taxes, subsidies, regulations, and customs dues were all criticized as "artificial" constraints on the natural processes of improvement.

This was a remarkably hopeful view of mankind, maintaining that all things worked together for good, and that public interest and self-interest could complement one another. Its optimism was especially well expressed by Joseph Priestley, a Unitarian minister and educationist who had for a time regularly encountered Watt and Boulton at the Lunar Society in Birmingham and who was related by marriage to John Wilkinson, the ironmaster. He is remembered today for his discoveries in chemistry: he identified a distinctive component of air that would support burning and that soon after came to be known as "oxygen." However, at the time his social philosophy was at least as influential as his science. Thus Priestly held it to be a "universal maxim, that the more liberty is given to everything which is in a state of growth, the more perfect it will become." And in a visionary and enthusiastic response to the earliest phase of the industrial revolution, he asserted that "all things . . . have of late years been in a quicker progress towards perfection than ever."[36]

James Watt had too gloomy a temperament to think like that. When his son, also called James, developed radical views and went to France soon after the start of the Revolution, he and friends in Manchester wrote advocating more moderate attitudes, but stressing that they too believed in the "improving" condition of "Society . . . in Europe."[37]

This and other letters also convey enthusiasm about technical matters, which were clearly seen as contributing to the improving state of society. Some correspondents energetically compared notes on subjects such as iron roof trusses, or experiments on bleaching cloth.[38] An industrialist's whole family was sometimes infected by this buoyant mood, and it is worth noticing that while formal accounts of scientific activities in these circles mention only men, enthusiasm for science was more widely shared. John Banks, whose public lectures and research on engines was mentioned earlier, acknowledged his wife's help, commenting that she was "very accurate in making experiments in (natural) philosophy and some branches of chemistry."[39] Similarly, Annie McGrigor, Watt's second wife, carried out a series of experiments on chlorine bleaching in collaboration with her father. He owned a dye works near Glasgow and was one of the first British industrialists to attempt the use of the new bleaching process in the 1780s. Watt had designed and built the equipment, but letters written by Annie indicate that she planned and carried out much of the detailed research.[40]

Utilitarianism and Industry

Optimism did not long survive the beginning of a new century. Many people were alarmed by the French Revolution, and the long years of war with France, ending only in 1815, further hardened attitudes. Meanwhile, the new industrial towns were becoming overcrowded and unsanitary, and it was also clear that working conditions in the factories left much to be desired. In 1819 Parliament passed a first Factory Act, which attempted to regulate the hours during which children could be employed in textile factories. But while some people now saw the need for government-imposed reforms such as this, industrialists defended their freedoms and rights in terms of an increasingly narrow version of Adam Smith's philosophy.

Some, however, remained hopeful about the progress of science and of industrial society, and among these were a group of people associated with Jeremy Bentham (1748–1832). Having grown to maturity in the second half of the eighteenth century, Bentham had been influenced by Joseph Priestley's glowing optimism, and his principle that "The good and happiness of the . . . majority of the members of any state, is the great standard by which every thing relating to that state must finally be determined."

In particular, Priestley had used the phrase "the greatest happiness of the greatest number," and Bentham made this his own, seeing it as "the only rational foundation, of all enactments in legislation and . . . conduct in private life."[41]

This left open the possibility of reform or corrective action if the unfettered growth of an industrial economy did not create the "greatest happiness." It allowed more scope for optimism if hoped-for benefits were not immediately achieved. And Bentham's later writings and influence were such that modern authors have described him as "the philosopher of the Industrial Revolution."[42]

But Bentham viewed the industrial revolution from a greater distance than the other men discussed in this chapter. He lived comfortably in London and had no experience of the enormous factories and the overcrowded workers' housing of Manchester. However, his brother, Samuel Bentham, was a naval architect and shipwright, and he knew William Strutt of Derby. So Jeremy had some contact with the world of industry and technology.

Bentham had been trained in law, and was an active campaigner for the reform of Britain's antiquated and corrupt legal system. But many of his attitudes suggest a laissez-faire rather than a reformer's approach to economic life, and in some ways his views were similar to those of other people mentioned here. Like them he thought that society developed according to laws of nature and had a natural structure. He tended to agree, too, that by letting nature take its course—and by letting individuals work for their own self-interest—improvement could be obtained.

But Bentham also made important modifications to the usual laissez-faire approach. It seemed to him that if one could understand the laws of nature as they affected society, the art of government could be improved. Instead of leaving "natural" social development to take its course, governments could use their knowledge to control or improve on what "nature" was already doing. The guiding principle in all this would be "the greatest happiness of the greatest number." This was the test by which the actions of governments, industrialists, landowners, and even private individuals could be judged.

"Utilitarian" is the word generally used to describe the philosophy of Bentham and his group, but Bentham himself preferred to say that his criterion was not one of "utility" but of "greatest happiness." In discussing science and what would today be called technology, Bentham said that their purpose should be the increase of human happiness. His attitude can be compared with the views of some of the seventeenth-century authors discussed in chapter 5 who held that knowledge should be used for the welfare of mankind. Bentham was, in fact, strongly influenced by one of the most important of these, Francis Bacon, whose views on the value of science might also be called "utilitarian."

Bacon had been a lawyer, and it was probably his writings on the reform of law that first attracted attention. But Bacon's views on the organization of knowledge also interested Bentham, who believed that the first responsibility of a government was to collect information. For example, governments should be regularly informed about the size of the population and how fast it was growing. The first census in Britain was taken in 1801, and Bentham's stepbrother, Charles Abbot, was involved in its organization. Bentham also called for registers of harvests, food prices, the flow of money, and industrial production. And he thought that, given the information, legislators would be in a position to create "positive well-being," or programs of social welfare aimed at abundance for all.

In this respect, Bentham's principles contrast sharply with the laissez-faire economic philosophy. But that does not mean that Bentham was keen for British governments to intervene more in running the economy or regulating industry. He regarded existing government operations as amateur and incompetent. Few officials knew much about the workings of the economy. They needed better information and skills to interpret it. Until this was achieved, governments would be wise to avoid interfering and practice laissez-faire.[43]

One merit of Bentham's philosophy was that, in accord with the more compassionate aspects of Francis Bacon's thinking, it acknowledged that the pursuit of economic self-interest was not a sufficient basis for social improvement. Modifications of the process of economic development to promote greater happiness were sometimes necessary and desirable. In that sense it was a philosophy that gave scope for social idealism in industrial development. However, efforts to promote the greatest happiness were usually conceived by Benthamites in a paternalistic way, and tended to discipline the people they were designed to help whether in the operation of schools, or in relief for the poor, or in the management of factories.

This aspect of Benthamite thinking undoubtedly helps to explain its appeal to manufacturers who were keen to improve the lot of their workers, but not at the cost of sacrificing the gains they had achieved in developing the factory system. The central purpose of the early factory owners, according to one persuasive commentator,[44] was to exert a more decisive *social control* over a recalcitrant labor force and so extract more work from them.

Arkwright's clear-sighted approach to the development of the factory system seems to reflect this attitude. The attraction of the factory for him was not just that spinning by machine was faster than older methods, but that a person working with a powered machine had to continue working without rest as long as the machine was running. The machine was thus a medium through which managers and supervisors could control the worker. The Strutt family, closely associated with Arkwright and operating mills on the same basis, must have shared his attitude to the control of the workforce. But they had wider interests than Arkwright, and stronger feelings about the social benefits industry should bring.

William Strutt of Derby became one of the most active practitioners of Benthamism from the 1790s until his death in 1830. His work in this respect is full of examples of the application of technology for human benefit, notably in the building of houses for his workers (figure 36), in the design of heating and ventilating systems, and in plans for the new Derby Infirmary,[45] completed in 1810 and probably the most advanced hospital building of its day.

Strutt had been experimenting with central heating for buildings since the early 1790s, because some form of heating was needed in the cotton mills he built, and this was a shared interest with the Benthams. For example, Jeremy Bentham wrote to him in 1794 reporting on a recent patent for heating by steam,[46] and also made inquiries in London on Strutt's behalf about copper tubes, which were needed for experiments.[47] Then Samuel Bentham was in correspondence with Strutt about a related subject[48] in 1805.

Strutt's work on heating and hospital design is a fine example of the use of knowledge, wealth, and technical skill for a compassionate purpose, very much in the spirit of Francis Bacon (chapter 5), and comparable with the public and charitable works most highly praised by Alexander Pope (chapter 6). In that respect, Strutt's work is part of a continuing theme in which social idealism influenced attitudes to technology. But as we have seen, William Strutt was very much associated with the early factory system, and inevitably with the use of factories as a means of social control. This is another aspect of Benthamism about which more will be said in the next chapter.

Figure 36
Ground-floor plan and section of a terrace house built in 1795 for workers from the Strutt cotton mills at Belper, Derbyshire. The provision of a kitchen, larder, and storeroom in addition to the living room on the ground floor represents a high standard of accommodation for factory workers in the 1790s. Probably designed by William Strutt, the plan typifies his ingenuity by the way it interlocks with its neighbor in the terrace, whose stairs and larder are at the back instead of the front.

(Author's drawing, based on a survey of 7, Long Row, Belper, with M. W. Barley, "Industrial monuments at Milford and Belper," *Archaeological Journal, 118* (1961), pp. 236–239).

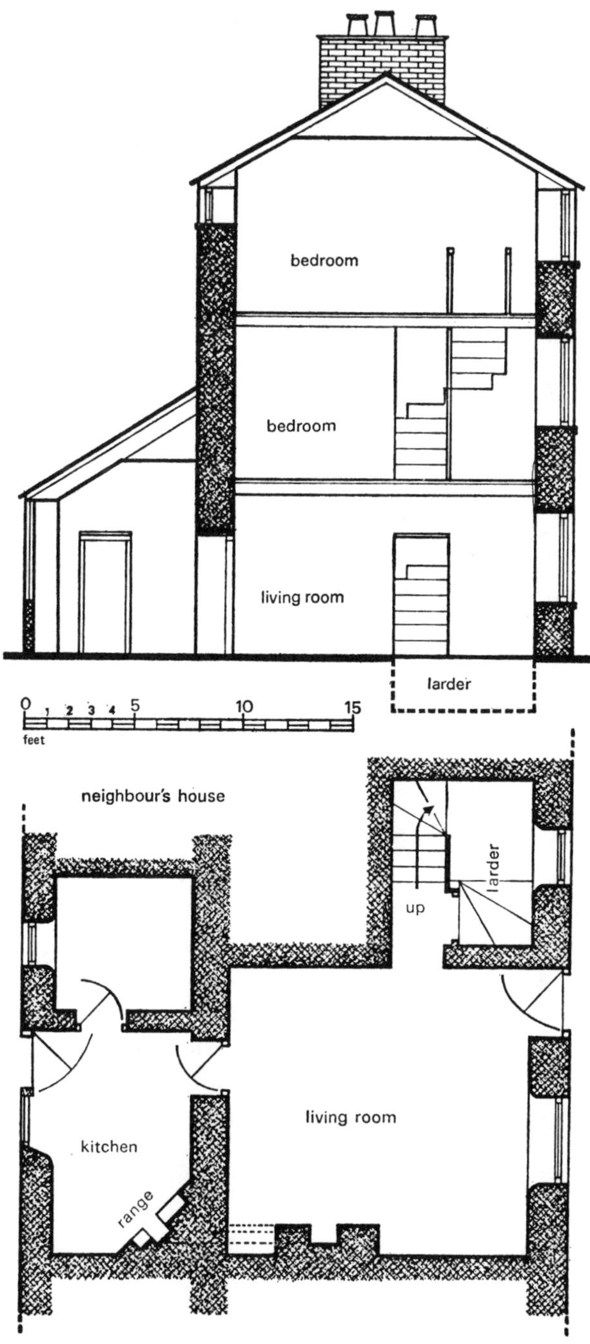
bedroom
bedroom
living room
larder
0 1 2 3 4 5 10 15
feet
neighbour's house
up
larder
living room
kitchen
range

8

Conflicting Ideals in Engineering: America and Britain, 1790–1870

Steam Transport Systems

The first major transport *system* to use steam engines as motive power was in the United States. It was based on the Mississippi River and had begun to take decisive shape by about 1815. But while North America's great rivers offered opportunities for communication, they were also daunting obstacles where they had to be crossed. So another characteristic of American technology around 1800 was innovation in bridge design, including Thomas Paine's trial cast-iron arch (1786), Theodore Burr's laminated timber arch with suspended roadway (1804), and James Finley's flat-deck suspension bridges, the earliest built in 1801 being the first of its kind in the world.[1]

Experiments were made using steam engines to propel boats from the 1780s, the most successful being by John Fitch. One of his boats, launched in 1790 and operated from Philadelphia, may have covered 2,000 miles (3,000 km) in trials and passenger-carrying journeys. However, the first successful passenger steamboat service was not inaugurated until 1807, and that was on the Hudson River between New York and Albany. The boat used was the *Clermont,* designed by Robert Fulton with paddle wheels driven by a 19-horsepower engine bought from Boulton and Watt in England. Fulton had learned from the earlier American experiments, but had also observed steamboat trials in Scotland in 1802, and had worked on the problem in France.

Fulton had also recognized the potential of the Mississippi, and in the winter of 1811–12 had another boat built—a vessel displacing 370 tons—and this was soon in service on the lower reaches of that river. Such was the demand for the new form of transport that within six years there were some seventeen steamboats operating on the Mississippi and its tributaries. This large number of boats indicates a transport *system* rather than just an isolated service. The pioneer nature of the enterprise is shown by the fact that in 1820 there were more steamers on the Mississippi, and of larger size, than in all the rest of the world. The sixty-nine steamboats operating in 1820 displaced in total some 13,890 tons, according to Louis C. Hunter, while elsewhere in the world it is hard to account for more than 10,000 tons of steam vessels.[2] Even in 1830, the Mississippi rivers had almost half the world's tonnage of steamboats, although the actual number operating on European rivers and coastal seas was greater.

It is understandable that steam engines were successfully applied to drive boats before they could be regularly used in road or rail vehicles, because the early engine occupied a large space in proportion to the power it produced. This was partly because the use of low-pressure steam meant that the engine's cylinder had to be relatively large, and partly also because of the method of driving the flywheel via a heavy beam above the engine. The earliest steam boats were large enough to accommodate all this, though in the *Clermont* the overhead beam was replaced by two shorter levers or beams, one on either side of the cylinder (figure 37). In some other boats on the Hudson River, an overhead beam was exposed to view above the cabin roof.

Robert Fulton, designer of the *Clermont,* was a professional artist rather than an engineer, and John Fitch had considerable artistic ability. The historian Brooke Hindle sees a significant affinity between art and engineering here, which he interprets mainly in terms of the ability to think visually that engineering often demands.[3] The same argument could, of course, apply to the "artist engineers" of the Renaissance (chapter 3). However, there is also the point that C. S. Smith has made in relation to metalworkers,[4] that aesthetic and technical motivations (or ideals) are often intertwined.

Robert Fulton's vision was more than just artistic, however. He could see enormous potential for the steam engine in river transport on all the great rivers of the world, not just the Mississippi. In 1812 he was writing letters to India about the potential for steamboats on the Ganges.

Another key innovator was Oliver Evans, who played a part in several early American steamboat projects and pioneered high-pressure engines. His first successful engine of this sort

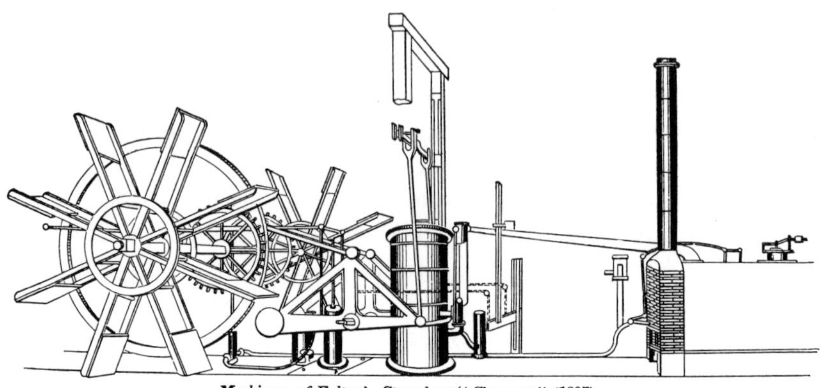
Machinery of Fulton's Steamboat "Clermont" (1807).

Figure 37
The engine and paddle wheels of Robert Fulton's steamboat *Clermont*. The engine's cylinder (center) has side-levers rather than an overhead beam. The connecting rods that drive the paddle wheels (left) via cranks (not seen) are attached to an arm whose bracing forms a triangle on the upper side of each side-lever.

(Source: Edward H. Knight, *The Practical Dictionary of Mechanics*, London: Cassell, no date, c.1890, 3 vols., Figure 3475. By courtesy of Ironbridge Gorge Museum Trust.)

was built in 1801, and in 1803 he supplied an engine for a steamboat experiment on the Mississippi. In 1811 Evans established a works for making steam engines at Pittsburgh, a city that would soon be a major center for building riverboats, and by 1814 he could claim that there were five engines of his manufacture at work in Pittsburgh industries, in flour mills, a paper mill, and at an ironworks. His work on flour mills was particularly significant, not only for the engines he supplied but also for the grain- and flour-handling system he invented that automatically ensured a smooth flow of material through the mill.

Although the Evans firm was too busy supplying industrial customers to build many engines for steamboats, other builders were adopting Oliver Evans's ideas, often infringing his patents. Many boat engines designed to operate at the low pressures associated with Robert Fulton were later modified to work at high pressures. Another innovation was the introduction in 1816 of an engine with a cylinder that was almost horizontal in a boat called the *Washington*, a 400-ton vessel. The engine drove the paddle wheels via a long connecting rod without either a flywheel or beam. All this saved space and achieved considerable mechanical simplification; the horizontal engine soon became standard for Mississippi steamboats.

From 1817 until the 1850s, steamboats on the Mississippi and its tributaries played a crucial role in allowing the economy of this part of America to develop. They also enabled settlers to reach relatively empty lands further west via tributaries such as the Missouri, Arkansas, and Red River.[5] In Europe, by contrast, there were no areas that could be opened up by riverboats in quite this way. A dense network of roads already existed, and in Britain most heavy freight went by canal or coastal shipping. Experiments with a steamboat on the Forth and Clyde Canal had shown up the disadvantages of steam power in these conditions. Boats driven by paddle wheels were too wide for many canals and caused erosion of the banks. Thus it was chiefly on the River Clyde in Scotland, not the connecting canal, where steamboats met with success. Ten of them were operating before 1817, whereas boats on canals continued to be hauled from the bank by horses.

Waterways of the latter kind became important in the United States following construction of the Erie Canal between 1817 and 1828. This was 360 miles (580 kilometers) in length and connected the Great Lakes with the Hudson River. It was built with knowledge of British canals and to the same small dimensions. This proved a disadvantage, and the locks and channel had to be reconstructed with more ample dimensions later.

The Coming of the Railroads

Meanwhile, in Britain, the year 1825 had seen the opening of the Stockton and Darlington Railway, which attracted attention to the potential of railways for long-distance transport. Prior to this date, railroads had been seen mainly as adjuncts to canal and river transport, consisting of very short lines with timber rails along which wagons traveled from quarries or mines to riverside wharfs. Such "railroads" (as they were called in Shropshire) or "wagonways" were thus not a transport system in their own right, but merely feeders for the river and canal system.

The first railroad locomotive had been built by Richard Trevithick, who pioneered high-pressure steam in Britain at just about the same time as Oliver Evans in the United States. This locomotive was completed in 1804, but it was not until 1812 that machines of this sort were in regular use, and these were at coal mines in the north of England. One reason for the delay was that there was difficulty in making a railroad track strong enough to take the weight of locomotives. Trevithick's first locomotives were of little use because of rail breakages.

By 1825, when the Stockton and Darlington Railway opened, these problems had been to a large extent solved, and although this line was still basically a mine railroad, carrying coal to a wharf on the River Tees, it did demonstrate the potential of railroads for carrying passengers and enabled locomotive haulage to be compared with horses (which were also used). The line attracted many visitors, including William Strickland, an architect and engineer who was there on behalf of a Pennsylvania society "for Promoting Railroad Improvements." It so impressed people that several more railroads were soon being built in Britain and on the continent of Europe. Some of these at first used horses for haulage, and it was not until George and Robert Stephenson built their famous locomotives *Lancashire Witch* and *Rocket* in 1828 and 1829 that severe limitations on speed and efficiency were overcome by more compact connecting rods and valve gear, and (in *Rocket*) by the multitube boiler. It was with locomotives having these features, and soon with horizontal cylinders also, that the first railroad in Britain to be designed for public transport between two cities opened in 1830. This was the Liverpool and Manchester Railway.

In America, the first comparable line was the Baltimore and Ohio, built between 1827 and 1830. It was planned with full knowledge of the Stockton and Darlington, using the same type of rails and the same gauge. Simultaneously, experience was being gained with locomotives on a line associated with the Delaware and Hudson Canal, and then in 1830, on a railroad being built from Albany to Schenectady. Locomotives imported from England were not a success and needed modification before they could cope with the sharp curves on the Albany line, and this led John B. Jervis to invent his four-wheel bogie.

These first imported locomotives came from workshops in northern England, opened in 1823 under the name of Robert Stephenson & Co. specifically to build railroad engines. In other places, however, workshops that had built machines for textile factories were adapted to build locomotives. For example, in America during the 1830s, locomotives were produced at Manchester, New Hampshire, near Lowell, Massachusetts, and in Philadelphia, all in connection with workshops that also produced textile machines.[6] The same evolution occurred in two or three machine shops in Manchester, England. Most notably, Richard Roberts, inventor of a "self-acting" or automatic spinning machine (or "mule") became a well-known locomotive builder (with his partners) under the name Sharp, Roberts.

A special feature of locomotives and spinning machinery manufactured under Roberts's supervision is that he aimed to have all their components made accurately to standard dimensions. This allowed for simplified assembly of the finished machines, because the components slotted into place without detailed adjustment and filing to shape. It also simplified maintenance, and engineers on the railroads were often pleasantly surprised to find how easily spare parts supplied by Sharp, Roberts would fit onto engines under repair.

While questions must be asked about whether all components of machines made by Roberts were fully *interchangeable,* his workmen used standard templates or jigs in cutting the shapes of individual parts, and it seems that dimensions were regularly checked using gauges. Roberts also designed "self-acting" machine tools that automatically disengaged when a component had been machined to the correct size. In 1835 a description of his works contained the following statement: "Where many counterparts or similar pieces enter into spinning apparatus, they are all made so perfectly identical in form and size, by the self-acting tools, such as the planing and key-groove cutting machines, that any one of them will at once fit into the position of any of its fellows in the general frame."[7]

This principle of standardized and interchangeable parts is often regarded as an American innovation, but Musson and Robinson have argued that the American pioneers "were preceded in the application of such methods by certain early British engineering firms"—including that of Richard Roberts.[8] Before this judgment is accepted, however, we should note the probable ancestry of the idea of using templates or jigs and purpose-built machines to make objects of standard dimensions. Richard Roberts had gained his early experience of machine-building by working for Henry Maudslay (1771–1831) in London. Maudslay had developed strong views on accurate workmanship and the manufacture of standardized products as a result of earlier experience. First, while making door locks in the 1790s, he had recognized the value of forming all the components to strictly controlled dimensions so that they would fit together easily without time-consuming adjustments. Second, he had been involved in the construction of a series of machines for use at Portsmouth Dockyard in manufacturing the wooden pulley blocks necessary in ships' rigging.[9]

This latter experience demonstrated how objects of standard size and shape could be manufactured mechanically by a sequence of machine tools, each specially adapted to perform one specific task in the several stages of manufacture. The idea for this sequence of "special-purpose" machine tools had originated in the mind of a Frenchman named Marc Brunel while he was working in America. Brunel had become an American citizen in 1796, but came to England three years later specifically to get his block-making machines built and used.[10] At the same time, standardization of heavy guns and gun carriages was being pursued in the British army. However, this was partly due to a French example.[11] In 1765, a "uniformity system" for the manufacture of guns had been proposed by General J. B. de Gribeauval as the basis for reform of arms manufacture in France, and a few muskets with interchangeable parts were soon being made there, as Thomas Jefferson found during a visit in 1785. He examined six guns and verified for himself that a component from one could be readily used to replace corresponding components on another.

Developments in America

At about the time when Marc Brunel was in America, designing his machines for block making, Eli Whitney contracted to manufacture 10,000 guns for the United States government. He talked about interchangeable parts and did much to establish the idea, but he seriously underestimated the difficulties. It was one thing for French craftsmen to make small numbers of guns with interchangeable parts, and quite another thing to manufacture guns in thousands. Thus it took Whitney eleven years, until 1809, to complete his contract, and even then the parts in his guns were not interchangeable.[12]

Bitter experience during the war of 1812 demonstrated the impossibility of repairing guns in the field if interchangeable spare parts were not available, and in 1813 a government contract with Simeon North specified interchangeability. An effort was also begun at the government armory at Springfield, Massachusetts, to achieve the necessary precision in manufacture.

Guns were still being made mainly by hand, and the key to standardization was seen to depend on disciplined procedures, first for manufacture of components using suitable jigs (templates) to achieve the standard shapes required, and second for using gauges to check measurements of the item being manufactured against a standard model.

Gradually, more machine tools were introduced. Simeon North invented a milling machine in 1816. Two years later, influenced by knowledge of Brunel's block-making machines, Blanchard devised a lathe for making wooden stocks for guns. Then, in 1820 Harpers Ferry, Virginia, became the center for development of interchangeable gun manufacture, partly at a government-owned armory there,[13] and partly at a factory set up by John H. Hall, inventor of a breech-loading rifle. Hall planned his factory so that all components of the rifle would be made by machine and to this end devised drop-forging equipment as well as lathes and milling machines. However, he also progressed toward making components interchangeable due to the fixtures he used with these machines. The issue here is that every time a workpiece was moved from one machine tool to another, there was an error in positioning and fixing it to the machine. This error could be minimized if there was one point on the workpiece which could be used in locating it. Hall designed special fixtures that would hold each workpiece very precisely in relation to such location points.

In the next two decades, many more machine tools were devised, and the American government's Ordnance Department promoted interchangeability of parts both in its own armories and through contracts with private firms. Some degree of interchangeability may have been achieved early in this process, but only at the cost of accepting guns with a very loose fit between parts. In the 1820s, Hall's rifles were a big improvement, but it was perhaps 1840 before rifles of reasonable quality were regularly made with fully interchangeable parts. Even then, this was being achieved by only a few firms outside the government armories, of which one of the most important was Robbins and Lawrence of Windsor, Vermont.

It will be clear that the American government expended considerable resources over a long time to reach this point. It is striking, therefore, that a small number of machine builders in Britain, notably Richard Roberts, seem to have begun interchangeable manufacture without such intensive development work.[14] However, Roberts's success in making locomotive parts interchangeable is not really comparable with the achievement of making the intricate mechanisms of rifles. It is also true that British achievements were backed by a longer period of development than appears, since they all owed much to Maudslay's work from the 1790s onward.

Joseph Whitworth, a younger man than Roberts, had worked with Maudslay toward the end of the latter's life and then set up a business in Manchester when Roberts's reputation there was at its height. Thus he had opportunities to learn from both men. Whitworth became known for promoting the idea of standard screw threads and also for his "self-acting" machine tools. Like many machines produced by the Maudslay school, their object was to utilize automatic operation to achieve precision in the shape and dimensions of objects being made.

In 1853 Whitworth and a colleague went to America as British representatives to the New York Crystal Palace exhibition. When this failed to open on time, they arranged visits to a number of factories and workshops. Whitworth found fault with many of the machine tools he saw (which were less solidly constructed than his own), but was impressed by wood-working machinery and commented particularly on the way in which factory owners sought to mechanize every process.[15]

The New York exhibition was an American response to the Great Exhibition held in London two years earlier. This had been housed in a remarkable building of iron and glass, itself made from prefabricated, standardized parts. One American exhibit there was Samuel Colt's stand, with its show of revolvers. Colt had been running a factory in the United States to make his patent revolver for several years, but the Great Exhibition generated sufficient interest for him to set up a factory in London.[16]

Quality and Mechanization

The 1850s were a time of rapid development in armaments for several European countries, accelerating during and after the Crimean War. Brass and iron cannon were being replaced by larger steel artillery pieces, with the Krupp works in Germany casting some of the biggest gun barrels by continuous pouring of crucible steel. In 1851, Krupp exhibited a steel casting of record size at the Great Exhibition. Five years later, Henry

Bessemer's ideas for a cheap, bulk steel-making technique originated while he was working on a new type of steel artillery piece.

Meanwhile, small arms in Britain were still produced by craft-based methods in numerous workshops, chiefly in the Birmingham area. Individually, each gun so made was of high quality, but dimensions were not strictly gauged and parts were not interchangeable. Thus repairs could not be quickly made by inserting spare parts, and damaged guns had to be returned to a gunsmith's workshop for repair. During the Napoleonic wars an enormous backlog of unrepaired guns built up, and similar problems occurred in the Crimea.

In 1854 the British Parliament set up a Select Committee on Small Arms to investigate this. Its hearings were associated with a visit to America by John Anderson, a government expert who was seeking equipment for a government factory at Enfield, near London, that would manufacture rifles with interchangeable parts. Anderson's most fruitful visits were to the Springfield Armory and the Robbins and Lawrence factory, and when the Enfield workshops opened in 1858, they were equipped with imported machines based on Robbins and Lawrence practice, and American workmen trained their British counterparts. Several other nations adopted American armory practice for small arms manufacture in the next few years, and Russia, Spain, Prussia, Sweden, Turkey, and Egypt all purchased American machinery for this purpose.[17]

The evidence heard by the British Select Committee demonstrated somewhat varying aims among engineers and arms manufacturers and showed that they were motivated by different *technological ideals.* For example, evidence given to the committee about Colt's revolvers showed that parts were not fully exchangeable. So, as the historian David Hounshell puts it, Colt (like Whitney before him) must be understood as a promoter rather than a practitioner of interchangeable manufacture. But the fact that people were promoting interchangeable parts suggests that it had become a significant *ideal,* and Hounshell confirms this when he quotes an American of Whitney's generation talking about interchangeability as "this grand objective." He goes on to say that the United States War Department officials found the idea "irresistible" because of its "pure rationalism."[18] The French ancestry of this ideal was indicated earlier, and it will be recalled that France was the home of other attempts to establish technology on a rational basis, including the greater use of mathematics in engineering and the analysis of machines in terms of systematic kinematics (chapter 7).

When Samuel Colt gave evidence to the Select Committee, it became clear that the reason why components of his revolvers were not easily interchangeable was that he was working toward another ideal which to some extent took priority. He expressed the view that "There is nothing that cannot be made by machinery." Thus in planning his factories, his main goal was mechanization, and he believed that uniformity and exchangeability would be achieved as a by-product of mechanical production.[19] Similarly, when Whitworth visited America, he was very impressed by the enthusiasm he found in many places for mechanized production, and by the use of special-purpose machine tools to achieve it. It is apparent, however, that *interchangeability* and *mechanization* were distinct ideals. Interchangeability could be achieved using hand methods if craftsmen shaped components against hardened steel jigs with gauges to check dimensions. Indeed, the interchangeable-parts guns made in France in the 1780s and the early experiments at the Springfield Armory depended largely on hand methods.

Thus one conflict of aims evident in Colt's operation was between interchangeability and mechanization as the ultimate ideal. Another conflict was related to the question of how much value should be placed on the *quality* of the finished product. It is said that in the 1820s, efforts to make interchangeable parts for guns achieved uniformity only to within 1⁄32 inch (slightly better than 1mm). Colt's revolvers were no doubt better than this, but there were still problems with badly fitting parts. Only in the 1880s was it possible to make components uniform to within a thousandth of an inch.[20]

Another American development in the 1850s was the manufacture of sewing machines. Some manufacturers aimed to build them with interchangeable parts, usually with the aid of technicians who had worked in gun factories. However, the most expensive and probably the best machines were produced by the Singer company. They retained a great deal of hand work and did not aim at intechangeability, because of the badly fitting parts that so often resulted. Instead, the priority was for high quality in each individual sewing machine made, which was recognized as a "European" rather than an American approach.[21] Later, the company needed to adopt more mechanization to achieve larger volumes of output. However, it took many years to reconcile the conflicting ideals of quality, interchangeability, and mechanized production, and only in 1881 was it claimed that this had been done.

Patterns of demand and the expansion of the market clearly influenced this development in sewing machine manufacture. But we should notice that market behavior in different countries is influenced by cultural factors that include attitudes to quality and standardization. In America, the railroad system provided a large market for builders of locomotives, and operating companies were willing to buy locomotives of standard design (figure 38). In Britain, each different company wanted its own distinctive types. Thus while Richard Roberts's practice of building locomotives with interchangeable parts did have influence (notably on John Ramsbottom of the famous Crewe works), few standard types of locomotive were built in quantity, and the advantages of interchangeability were hardly realized.

Visions of Automation

The limited enthusiasm for mechanization in Britain as compared with America is somewhat surprising when one considers how thoroughly Arkwright had been committed to mechanizing cotton spinning, and how his influence had spread through the rest of the textile industries. After Arkwright's time, the development of self-acting or automatic looms and spinning mules was a powerful stimulus for some people in Britain, encouraging thought about how far systems of mechanization could advance. One such person was Andrew Ure, a former student of medicine and teacher of chemistry in Glasgow, who had moved to London to work as a writer and a consultant in applied chemistry. One book he compiled was a *Dictionary of Arts, Manufactures and Mines*. In this, the word "automatic" was defined as the "economic arts . . . carried on by self-acting machinery," a sentence that was amplified by reference to the meaning of another word: "Manufacture," in its original sense, meant work carried out by hand. But "in the vicissitude of language" it had come to mean almost the exact opposite, for the ideal manufacture "is made by machinery, with little or no aid of the human hand, so that the most perfect manufacture is that which dispenses entirely with manual labor."

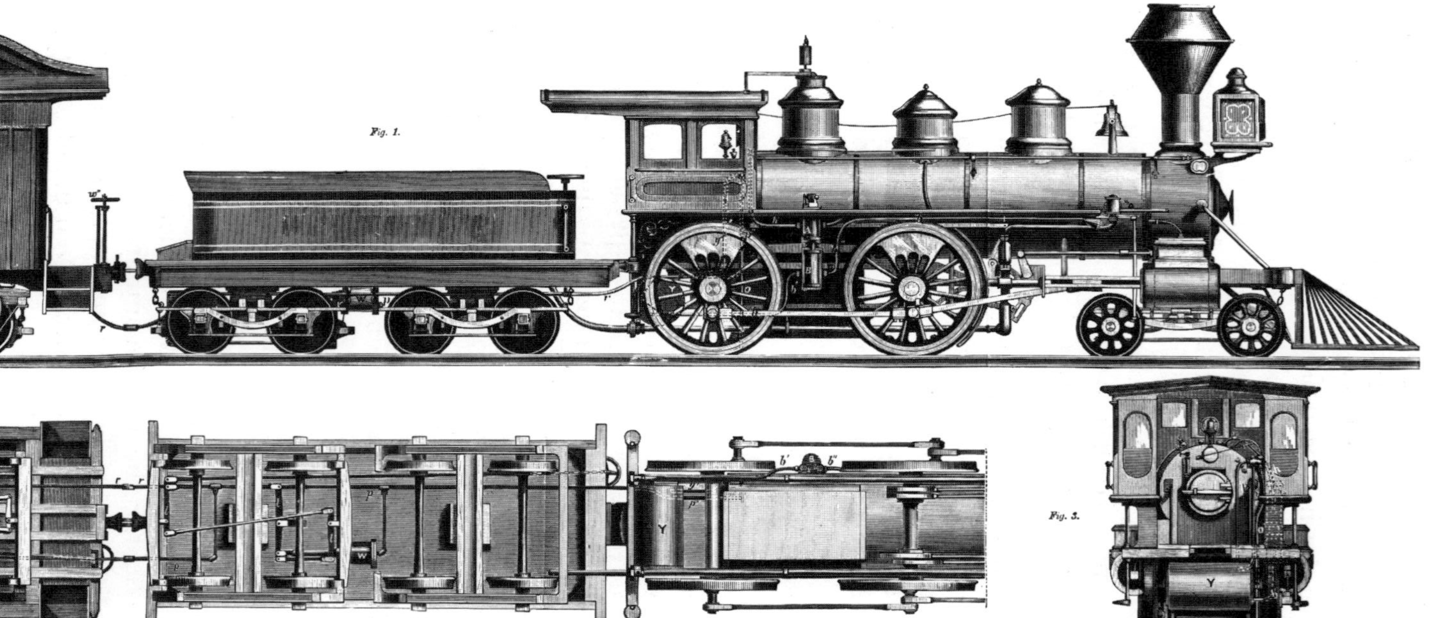

Figure 38
Locomotives such as this with four coupled driving wheels and a four-wheel bogie were very widely used in the United States, making standardization of parts feasible and economically attractive. In Britain, there was a greater variety of locomotive types, reflecting the preferences of individual companies, and little standardization was possible. The illustration is designed to show the braking system invented by George Westinghouse toward the end of the 1860s. This entailed running a continuous air pipe, marked *r,* along the length of the train.

(Source: Edward H. Knight, *The Practical Dictionary of Mechanics,* London: Cassel, no date, c.1890, 3 vols. By courtesy of Ironbridge Gorge Museum Trust.)

To such a degree had this idea captured Ure's imagination that he made it almost the basic axiom of his *Philosophy of Manufactures,* published in 1835: "The most perfect manufacture is that which dispenses entirely with manual labour. . . . The philosophy of manufactures is therefore an exposition of the general principles, on which productive industry should be conducted by self-acting machines." The ideal factory, therefore, was "a vast automaton, composed of various mechanical and intellectual organs, acting in uninterrupted concert for the production of a common object, all of them being subordinated to a self-regulated moving force." That moving force was of course the steam engine.

It is clear that Ure's vision of an automatic factory resulted from the great impression made upon him by the self-acting machines in the textile mills: "It is in our modern cotton and flax mills that automatic operations are displayed to most advantage; for there the elemental powers have been made to animate millions of complex organs, imparting to forms of wood, iron, and brass, an intelligent agency. And as the philosophy of the fine arts, poetry, painting, and music, may be best studied in their individual master-pieces, so may the philosophy of manufactures in these its noblest creations."[22]

Ure had heard of a series of patents taken out by J. G. Bodmer in 1824 for use in cotton spinning mills. The effect of Bodmer's inventions was to allow the output of one machine to be fed directly into another for the next stage in the process, without human intervention. Thus "the several organs of a spinning factory are united into one self-acting and self-supplying body—a system most truly *automatic.*"[23] A fully automated factory was not a real possibility in Ure's time, but in these ways he was able to imagine what it would be like. It was, he said, the "constant aim and tendency of every improvement in machinery to supersede human labour altogether."

If a particular process could not be completely mechanized, Ure said, it should be organized according to the division of labor, with complex tasks broken down into a series of very simple operations, each of which should be done by a separate worker. Wherever possible, machines or special tools should be introduced to make these elementary operations even simpler, so that less skill was needed. The aim was to diminish the cost of the necessary labor, by "substituting the industry of women and children for that of men; or that of ordinary labourers, for trained artisans."[24]

Ure recognized no moral objections to the employment of children, and claimed that they actually enjoyed working in factories: "the work of these lively elves seemed to resemble a sport." As for the other workers, he thought that the effect of introducing automatic machines was beneficial, or as he put it, "philanthropic, as they tend to relieve the workman either from niceties of adjustment, which exhaust his mind and fatigue his eyes, or from painful repetition of effort, which distort and wear out his frame."[25] Ure's views are, indeed, a complete contrast to those of his contemporary, Robert Owen, who until 1824 was manager and part-owner of the New Lanark cotton mills. Owen disliked the way in which skilled manual work was being replaced by mechanism; and the division of labor for him meant division of interest, so that every job became totally uninteresting. This was wasteful in human terms, and led to frustration, boredom, and crime.

A less damning analysis of the manufacturing system, which at the same time was more sensitive than Ure's, was given by Charles Babbage in 1832 in his classic work *On the Economy of Machinery and Manufactures.* Babbage stressed the division of labor in the organization of industry, and he held that "Governments ought to interfere as little as possible between workmen and their employers." However, he did agree that the government should limit the employment of children in factories, arguing that this was a matter of giving protection to the weaker party.

Babbage made his own contribution to the development of automatic techniques. In 1824 he consulted Peter Ewart about looms, and in particular about devices that would stop the working of such a loom in the event of a warp thread breaking.[26] His interest in automatic looms extended to work that had been done in France by Jacques de Vaucanson (1709–1782) and J. M. Jacquard (1752–1834). Vaucanson had a reputation for making "automata": that is, animated clockwork toys, but he had also devised a silk-weaving loom in which the pattern woven into the fabric was produced by the operation of an automatic gadget. In Jacquard's improved version of this loom, the pattern of the cloth was determined by a series of cards with holes in them that were fed into the machine. A group of metal rods were pushed against each card; those that passed through a hole were lifted several inches, and this controlled the rearrangement of warp threads between each passage of the shuttle.

The device was conceived for use with handlooms in France, but greater advantage was gained from its application to power looms in Britain. In addition, it provided Babbage with the idea of punched cards, and his experiments with mechanical computers probably owed much to the example of this mechanism. From about 1823, Babbage became deeply involved with a project to construct a computing machine, and he employed Maudslay's associate Joseph Clement to help him. Clement was expert in the design and construction of machine tools and was especially skilled in any branch of mechanical construction that called for great precision and accuracy.

Since Babbage had so direct an interest in the hardware of automation, it is ironic that the possibility of automated industry was much more clearly expressed by Andrew Ure, in his book of 1835. In the next three decades, Ure's remarks on self-acting machines influenced several people who wrote on the effects of mechanization, including most notably Karl Marx.[27]

It is worth noting one issue that links the arguments about mechanization put forward by Ure, Babbage, and Marx. All three authors saw that fewer craftsmen of the old sort were needed in industry, because the skill was now being built into "self-acting" machines. Ure claimed that working people were glad to avoid the burden of taking pains over their work. But Marx argued that the simplification of work by machines "is used to make workers out of those who are just growing up, who are still immature, *children,* while the worker himself has become a child deprived of all care. Machinery is adapted to the weakness of the human being, in order to turn the weak human being into a machine."[28] Some historians also note that mechanization was often introduced in response to strikes as a means of making a company less dependent on unruly elements in the labor force.[29]

However, in order to understand why automation was regarded with such enthusiasm by Ure and others, we may usefully look beyond the issue of social control of labor and notice that from quite early in human history, automatic gadgets have fascinated inventors. In the Greco-Roman world, devices were invented to open temple doors without human intervention. Hero of Alexandria wrote about these and other automata. In the Islamic civilization and in the European Middle Ages also, clocks were equipped with model human figures that struck bells on the hour. Jacques de Vaucanson, the French technician whose automatic loom was a forerunner of the Jacquared loom, was a maker of automata and clockwork toys in precisely this tradition. It is not too much to claim, therefore, that long before automatic machinery became a technical ideal for factory owners, the automaton was an object of interest and fantasy for many technicians. One commentator sees this as not just a simple fascination with a mechanical device, but as a perverse interest in the "ideal of a world without people." He describes it as the "fantasy of a perfectly ordered universe . . . controlled (and) unpredictable," which is an attitude expressed clearly in many of Andrew Ure's remarks.[30]

Moral Inventions

Achieving interchangeable parts depended on disciplined work as well as on jigs and fixtures, and in America this was expected to "instill values conducive to the moral growth and well-being of the country."[31]

There are parallels here with the philosophy of Jeremy Bentham, mentioned in the previous chapter. Bentham had genuine social ideals oriented toward the "greatest happiness" of people. But he was consistently attracted to forms of organization and technology that would bring about better behavior. Artificial devices that had this result he referred to as "moral inventions." Examples included a factory built by William Strutt in 1813 that had a circular plan so that the workers could be more efficiently supervised from a central control point. It was based on an idea of Bentham's, which he called a "Panopticon." Another moral invention of Strutt's was a toilet compartment in which the user's action in opening the door to leave the toilet automatically flushed the water closet, thereby compensating for forgetfulness.[32]

Sanitation later became an important field for the application of a Benthamite approach to technology due to the work of Edwin Chadwick, who had acted as Bentham's secretary and in 1832 nursed him through his last illness. Chadwick's fame rests on his report: *The Sanitary Conditions of the Labouring Population of Great Britain,* written in 1842. This argued that an important reason for the filthy state of most British cities was the high cost of removing decomposing refuse and human excreta by existing methods, which consisted mainly of carting it away, often for use as manure. Such expense "may be reduced to one-twentieth or to one-thirtieth, or rendered inconsiderable by the use of water and self-acting means of removal by improved and cheaper sewers and drains."[33]

This implied that houses should be equipped with water closets and that wastes should be removed as a suspension in water. If a constant supply of water were available in every street—which was rarely the case in Britain—and if every water closet (or other latrine) were connected to a sewer, wastes would be removed immediately. What Chadwick was advocating, then, was a "combined system" of water supply and sewerage. If properly designed, this would be a "self-acting" system that would automatically ensure that no accumulations of waste built up close to people's homes. It was a "moral invention" also, in that it encouraged hygienic behavior.

Chadwick's use of the term "self-acting" is significant in a more restricted and technical sense also. He had taken evidence from an engineer named John Roe, who had realized that if sewers were laid to a carefully calculated and consistent gradient with smoothly rounded bends, they could be made self-cleaning. That is, the flow of water would always be sufficient to carry away the solid component of the sewage, and no sediment would be deposited. Roe also showed that a self-cleaning sewer could be much smaller than most existing drains, partly because it would pass the flow of sewage more efficiently, but also because it was no longer necessary to design on the assumption that men would have to crawl into the sewers to clean them out. So house drains and branch sewers could now be laid as glazed earthenware pipes rather than being built as small, brick-lined tunnels.

In drawing together these ideas in his report, Chadwick established comprehensive principles for the development of urban sanitation in Britain. Most historical accounts describe subsequent events in terms of a series of engineering triumphs implementing Chadwick's ideas, but this is almost the reverse of the truth.

An Act of Parliament based largely on Chadwick's recommendations was passed in 1848, and this established a General Board of Health. In the next five years, many small towns introduced plans for improved water and sewerage systems in collaboration with the Board of Health, and here one certainly can see Chadwick's ideas being applied. However, a number of major cities already had active sewer-building work in hand (for example, Manchester),[34] or were taking their own initiatives to improve water supplies (Nottingham).[35] These cities opted to remain independent of the Board of Health and were relatively little influenced by Chadwick.

London was covered by separate legislation, and it was here that controversy raged most fiercely, with Chadwick often tactless and uncompromising, and many in the engineering profession questioning his competence. There were also many points of technical disagreement, one about the principle of small-bore, self-cleaning sewer pipes. An engineer named Bazalgette arranged for 122 of these pipes to be dug up and examined, often "secretly, and after night fall."[36] He then wrote a damning report that was widely circulated, claiming that 23 of these pipes were cracked and 66 were "chocked." This was a biased sample, however, and within a few years engineers who had once criticized pipe sewers were regularly installing them because they cost one-third of the alternative, and they worked.

Another dispute was about the very large main sewers needed in London. This rumbled on for several years, and by the time it was settled, the original Board of Health had been dissolved and Chadwick had lost his job. In 1856 a new Metropolitan Board of Works was set up, and Joseph Bazalgette, Chadwick's arch-critic, was appointed its chief engineer. The plan for the main drains was now that they should discharge into two enormous intercepting sewers running parallel with both banks of the River Thames. These would carry the sewage downstream, and beyond the confines of the city it would pass into settlement tanks from which effluent would be pumped into the river estuary on the ebb tide. The enormous pumps necessary for this task were driven by steam engines that were not only impressive in size but richly ornamented in their construction.

Engineering works for water and sewerage schemes in other parts of Britain were also carried out in a heavily monumental style, to match the aqueducts of Rome, one engineer claimed of his Glasgow project.[37] Other pumping engines were also highly decorated, in one case near Nottingham with a smokestack disguised as an Italian campanile. Whiteacre pumping station in Staffordshire had Egyptian-style columns oddly mixed with Gothic detail, and around the top of the engine's enormous cylinders, gilded eagles served as brackets to support an inspection platform.[38]

The engineers who built these fantastic monuments generally gave low priority to the matter about which Chadwick felt most urgency, that is, measures to remove waste from the vicinity of people's homes. So although improved sewers were laid and clean water supplies were developed on an impressive scale, many householders had to continue using very primitive latrines in their backyards with inefficient systems for removing excreta from them. The advantage for health to be gained by full implementation of the "combined system" were thus delayed. One can sympathize with Chadwick's biographer, who accuses the leading engineers of the 1840s and 1850s of "a swashbuckling disdain for the social evils around them."[39]

Railroads and Bridges

Many of the engineers with whom Chadwick was in conflict had gained the greater part of their experience building railroads. They included Robert Stephenson who, with his father George, had built the first intercity lines in Britain. Stephenson's policy in planning these railroads was to use embankments, cuttings, and tunnels when necessary to avoid steep gradients and sharp curves. In America, by contrast, where there were much greater distances to cover, railways were built more quickly and with a smaller labor force, so heavy earthworks were avoided wherever possible, even if this meant unfavorable gradients and an irregular curving route.

Another factor in early British railroad developments was their relationship with innovations in the iron industry, especially those that made it possible to produce the first fully satisfactory wrought-iron rails in 1820 or 1821 (at Bedlington Ironworks in Northumberland). In other countries where it was desired to build railroads, the first rails were often imported from Britain, but rail-rolling mills were soon established in Belgium, then later in Germany and France.

The situation in the United States was different since the industrial base was initially smaller than in France or Britain, but the rate of building railroads was faster. Thus the United States was soon taking a high proportion of British exports of rails, and the prosperity of the British iron industry became closely linked with American railroad construction. Indeed, by supplying rails on favorable terms, some British ironmasters became important investors in American lines. It is instructive to observe that in the 1850s, while 3,000 miles (5,000 km) of railroad were built in Britain, about the same in Germany, and slightly more in France, *six* times that route mileage was opened in the United States. This decade included the American railroad boom that began in 1854, but railroad mileage had been extending faster there than anywhere else for almost two decades. American railroads were usually single rather than double track and took a lighter type of rail than was used in Britain, but despite this, the amount of pig iron consumed in British ironworks between 1852 and 1859 to manufacture rails for the United States has been estimated to have averaged 234,000 tons per annum, by comparison with 174,000 tons needed by Britain's own railroads.[40]

With an economy so dynamic in every other way, it would have been surprising if Americans had neglected the manufacture of their own railroad material. As already mentioned, locomotives were often built in places where a mechanized textile industry had developed, sometimes in workshops that had formerly built machinery for cotton-spinning mills. Hardly any locomotives were imported after 1836, and in 1839 the majority of the 450 locomotives in the United States had been manufactured within the country.

The production of rails in any quantity was a more difficult proposition, however. It depended much more on bulk iron production, and hence on developments in the iron industry that were only just getting under way in America around 1830. Some of the changes had to do with using coke instead of charcoal, and some were to do with adapting established types of furnaces to suit the anthracite coal that was plentiful in Pennsylvania. Progress was rapid, so that while 80 percent of rails came from Britain during the 1840s, domestic production of iron rails overtook imports in 1856.

Just as earthworks and rails were often lighter in America than in Britain, so also railroad bridges were often less substantially built and were more often made of timber.[41] Some bridge trusses were designed to have timber for members that were in compression, with iron rods used as tension members. Designs were rethought when iron was more widely used, and in 1844 Thomas W. Pratt and his father Caleb patented the very economical "Pratt truss." One example of its use was for a railroad viaduct crossing the Delaware River at Trenton, New Jersey.

Suspension bridges with wire cables were pioneered in France, mostly for highway traffic, by Marc Seguin and Louis-Joseph Vicat. However, the fullest development of wire suspension cables must be credited to John Augustus Roebling who, in 1844, developed a method of spinning such cables on site from continuous wire. His first application of this technique was for a suspended aqueduct on the Pennsylvania Canal. In 1851 he began construction of a suspension bridge over the Niagara Gorge that carried both a highway and a railroad. However, his greatest achievement was the Brooklyn Bridge, which carried street traffic over the East River in New York. It was a suspension bridge with diagonal stays, and its main span was 1595 feet (486 meters). That was 50 percent longer than any bridge then existing. Sadly, Roebling died after an accident on the site shortly before construction began in 1869, but the job was taken on by his son, Washington A. Roebling, who in turn became ill as a result of an accident, and whose wife then played an important role in ensuring that construction continued. When spinning of the cables began in 1876, this became the first suspension bridge to be made with cables of steel rather than iron wire.

Symbols and Conflicting Ideals

Interpretation of the significance of the Brooklyn Bridge has been a matter of minor controversy among historians of technology. Some have seen it as "more than a structure to span a river; it is a symbol of victory, of man's conquest and achievement." Others describe it as a "work of passion and discipline," to be compared with the cathedrals of thirteenth-century Europe because of its "symbolic meaning."[42]

This talk of the symbolism of the Brooklyn Bridge has been strongly criticized by C. W. Condit, a distinguished historian of structural engineering and architecture. He argues that while the cathedrals symbolized a world view that was made fully explicit at the time they were built, there was no stated intention that the Brooklyn Bridge should convey any comparable meaning. Thus one cannot claim any deliberate symbolism for it.[43]

But symbolism is not always explicit or consciously intended. If we say that a bridge is a symbol, all that we are saying is that it *means* something beyond its practical function, perhaps just to its builders, or perhaps more particularly to the public at large. Cossons and Trinder have demonstrated that the pioneer iron bridge near Coalbrookdale in Shropshire meant a good deal to contemporaries by noting how frequently artists painted it, and how often images of it were used to decorate household objects.[44]

With the Brooklyn Bridge, visual images dating from soon after its completion are plentiful (figure 39), and other evidence is provided by a striking speech made at the opening ceremony in May 1883 by Abram S. Hewitt, a local dignitary who later became mayor of New York. He compared the bridge with the pyramids of Egypt (a frequent point of reference when people think that engineers have created important symbols). He also said of the bridge that "when the airy outline of its curves . . . pendent between its massive towers . . . is contrasted with the overreaching vault of heaven above and the ever-moving flood of waters beneath, (which are) the work of omnipotent power, we are irresistibly moved to exclaim: 'What hath man wrought!' "[45]

As we saw in chapter 2, the explicit symbolism of the thirteenth-century cathedrals was that they captured something of a divine creation, the New Jerusalem, in their construction. The symbolism of medieval clocks was that they represented, even rivaled, the movements of the sun and stars in their perpetual and regular motions. Now we are told that the Brooklyn Bridge has something of the vault of heaven about it, indicating that for one citizen of New York at least, the bridge had much the same kind of symbolism as the cathedrals. Indeed, a recurring theme in the symbolism associated with technology seems to be the idea of human invention as mirroring and even rivaling nature's (and God's) creative energy. While automata, discussed earlier in this chapter, achieved an uncanny imitation of living forms, a bridge, a cathedral, or a pyramid represents inanimate cosmic structure.

At this point it is worth recalling the ornamentation of large pumping engines at water and sewage works in Britain. Steam engines in British textile factories also featured classical decorative detail, and some were placed in ornate engine houses. Almost always in Britain, where steam engines were used, even if they were devoid of decoration, they were regularly cleaned and polished and sometimes were given personal names.

By contrast, the steam engines that powered Mississippi riverboats were rarely polished and never ornamented, and maintenance was so neglected that some boats were a danger to their passengers. This is despite the decorative superstructure of later

Figure 39
Brooklyn Bridge, New York, built between 1869 and 1883. The illustration probably dates from the 1880s, and uses a slightly distorted perspective to emphasize the cables associated with the nearer tower. Apart from the usual suspension cables, there is also a series of diagonal stays.

(Source: Edward H. Knight, *The Practical Dictionary of Mechanics*, London: Cassel, no date, c.1890. By courtesy of Ironbridge Gorge Museum Trust.)

boats, which served to advertise their luxury and comfort. Noting all this, Hunter adds that riverboats were flimsily built for a short working life because of the American belief "that improvements will be discovered in everything." Boat-builders explicitly said that there was no point in having vessels that would "last too long because the art of steam navigation was making daily progress."[46] Similarly, American machine tools were of light construction compared with the solidly built equipment that British engineers designed. So while British steam engines and machine tools were treated as *monuments* symbolizing progress already achieved, the whole ethos surrounding their American counterparts was one of expectation of future progress.

Similar contrasts can also be observed in civil engineering. The boldest project in Britain during the early "railway age" was the Britannia Bridge, which carried the Holyhead railroad over the Menai Straits in North Wales. Built under the direction of Robert Stephenson between 1845 and 1850, the bridge was a wrought-iron tubular girder (or box girder) with 460 foot (140 meter) spans. It was a pioneering structure that established, for the first time in Britain, techniques for building large wrought-iron girder bridges. But it was a stronger, heavier, and more costly work than strictly necessary, and a century later was carrying loads far larger than those for which it was designed. When Stephenson heard of Roebling's Niagara Gorge bridge, he was appalled by its lightness and by the use of a suspension bridge for a railroad. In a letter to Roebling, he commented, "If your bridge succeeds, mine is a magnificent blunder."[47]

One can begin to understand, in this context, why some engineers were so hostile to Chadwick's low-cost, small-bore pipe sewers. The pipes hardly matched up to the British tradition of solid, high-quality construction. At the time of the disputes about Chadwick's plans for London, Stephenson became a member of the Metropolitan Commission for Sewers and allowed his prestige to be used against Chadwick. He was particularly contemptuous of the idea of small-bore pipes, and said he "would not touch one." He "hated the very name of them."[48]

With respect of both bridges and sewers, therefore, Stephenson's attitudes, like those of many colleagues, tended to stress monumental construction and quality. In comparing this attitude to American views, we might say of the British that they tended to judge achievement and value quality in relation to individual products—guns, machines, steam engines, bridges. Americans, by contrast, seem to have been more concerned with *systems* for production, or for transportation on a continental scale. S. B. Saul characterizes the British attitude as "love of the technical product rather than the technique of production."[49] And while Britain was building railroads to consolidate earlier industrial achievements, America was still expanding into new areas and pushing back its western frontier. So while a suitable analogy for railway building in Britain might be the aqueducts of ancient Rome (some viaducts present a close visual parallel), one analogy that has been used with respect to American railroads is the space race of the 1960s, because the railroads were concerned with a conquest of geographical space rather as the 1960s adventure was a conquest of cosmic space.[50]

There are, of course, important economic factors that influenced the different directions taken by American and British technology. They include such things as relative wage rates of skilled and unskilled labor, the low wages paid to laborers and factory workers in Britain, and the consequent lack of a mass consumer market in Britain. But as the more discerning economic historians recognize, these economic influences do not account for all the observed facts, and we need also to notice the different cultures of a growing nation with an expanding frontier as compared with an increasingly complacent imperial power.[51]

9

Institutionalizing Technical Ideals, 1820–1920

Engineers and Government

Few would doubt the importance of economic incentives and market demand as influences on invention, innovation, and the development of technology. Equally, however, one may observe some engineers and technologists who seem to defy economic rationality in the way they work. Thus although the heavier engineering works and easier curves on British railroads as compared with American lines can be explained chiefly by reference to vastly different economic and geographical conditions in the two countries, one must also take note of the point made in the previous chapter about British engineers building monuments to progress, while Americans built with anticipation of advances still to come. There were certainly many British engineers who took a realistic view of the economic prospects of the railroads they planned, but there were some who used their high reputation and partial monopoly of expertise to demand that sufficient capital be raised to pay for engineering works of an idealized quality. Isambard Kingdom Brunel was an exponent of idealistic engineering if ever there was one, insisting not only on good alignments and easy gradients on his Great Western Railway (begun in 1835), but also adopting a broad gauge that was theoretically advantageous but not very practical. Later, he persuaded shareholders and businessmen to finance extraordinary ship-building projects, and because he invested personally in all his own schemes, he lost a good deal of money.[1]

It is clear, then, that economic pressures and market conditions did not wholly exclude the influence of idealistic or visionary impulses in technology. However, engineers who felt driven by such impulses sometimes had greater opportunities within public institutions that were to some extent insulated from the strict disciplines of market forces. In Britain, as was shown in the previous chapter, engineers working on municipal sewerage and water projects or employed by the Metropolitan Board of Works in London built monumental pumping stations and other structures of extravagently high quality. In the United States, the ideal of interchangeable parts in handguns was achieved largely because of work done in government armories (and by private firms with governmental contracts).

Thus to account for differences in style between American and British technology in the nineteenth century, one must look beyond economic conditions to consider a number of institutional factors. These include the status of professional engineers and the role of governmental and public institutions (such as armories). But within private companies also there were institutional developments that allowed idealistic objectives of various kinds to be pursued, sometimes for reasons of prestige and reputation.[2] Examples include monumental buildings and bridges built by some railroad companies, and firms that pursued ideals such as product quality (Singer, see chapter 8) or mechanization (Colt) beyond simple considerations of financial gain.

It can be reasonably claimed, then, that institutions of several kinds developed during the nineteenth century in ways that left considerable scope for "privileged technologies"[3] to escape the full rigor of market disciplines, and enabled idealistic patterns of innovation to evolve. Since everything had ultimately to be paid for, economic conditions set limits to how far technical ideals could be pursued, but idealistic impulses had a greater role than is usually recognized.[4]

This theme becomes clearer when international comparisons are extended to include France, where the *Corps des Ponts et Chaussées* had been set up in the eighteenth century as a government service for making maps, planning roads, and designing bridges (chapter 6). In 1820 the director-general of the *Corps,* Louis Becquey, published a general plan for developing canal transport throughout France that met with government approval, so that funds for as much as 2,000 kilometers (1,200 miles) of waterway were allocated. Meanwhile, in 1823 Becquey began to encourage the construction of railroads in France to perform the same function as many early lines in England, namely to link coal mines to canal or river transport. Thus in 1826 the Seguin brothers were granted a concession to build a coal-carrying railroad from Lyon to Saint Étienne. Marc Seguin was mentioned earlier as a pioneer of suspension bridges with iron-wire cables. Now he and his brother imported a first locomotive from England and increased its steam production fourfold by installing a multitube boiler. This was a device they had previously used in a steam boat on the River Rhône, ahead of its adoption by the Stephensons in Britain (chapter 8).

After a period of political upheaval in France that temporarily threatened the position of the *Corps des Ponts,* the planning of long-distance railroads began in 1832 under the direction of Victor Legrand. The basic pattern was to be a series of lines radiating out from Paris. However, the railroads were not conceived as displacing canal transport but rather as complementing it. Canals were regarded as best for heavy freight, and their construction was continued, often parallel with railroads, which were to provide for the rapid conveyance of passengers on many of the same routes.

The head of the railroad commission was Louis Navier, the author of a mathematical theory of the suspension bridge that offered a rational analysis of what Navier saw as a crudely empirical Anglo-Saxon invention. With the railroad, as with the suspension bridge, he "sought the ideal system."[5] He believed that heavy investments in civil engineering works were justified if they led to fast and safe travel. He thereby took a position opposite to that of the American builders of what he thought were cheap and dangerous railroads. Navier's aim of "solidity" and "beauty" in railroad construction seemed fine to some, but there were also people in France who could see that the rationally planned canal system had involved enormously extravagent expenditures. They feared that railroad planning was going in the same direction and in 1838 secured the defeat of Legrand's plan in the French parliament. Although this was only a temporary setback for the planners, its immediate consequence was that the Paris-Rouen line was built by a private company, which brought in a British engineer, Joseph Locke, who had a reputation for building cheaply.

Employment of inadequately trained foreign engineers was not allowed again, however, and French railroads were built very largely to the standards set by the *Corps des Ponts.* People were aware that Belgians, Germans, and Americans were building railroads with single track, sharp curves, steep gradients, and wooden bridges, but France did not want a "makeshift" system and adhered to the ideal of building "for posterity, not profit."[6] Many lines were, in fact, built by private companies and did make a profit, but it is also clear that French engineers, like their British counterparts, had a position that enabled them to build in an idealized manner. Moreover, they could go much further than in Britain in planning the railroad and canal systems as integrated networks.

Engineers working for private companies in France were often critical of official plans and claimed that cost-conscious design could save millions of francs. They organized themselves into a society of private engineers, calling for "English-style industrial liberty,"[7] and demanding the virtual dissolution of the *Corps des Ponts.* But although there were phases of liberalism in politics that favored such demands, the *Corps* was too well established for its role to be easily changed, and its ideals strongly influenced the private sector. Thus in the 1850s and 1860s elegant masonry arch bridges of more adventurous design than British railroad viaducts were sometimes the work of *Corps* engineers, while important refinements in the theory and design of iron railroad bridges were developed by the private engineer Gustave Eiffel. And while the private engineers regularly argued against the extravagant use of funds in official projects, Eiffel was responsible, using his own money, for the most extravagantly useless project of all, the Eiffel Tower. Built for the Paris exhibition of 1889, the tower was significant mainly for what it symbolized about "secular faith" in technical progress.[8] It incorporated the results of its designer's pioneering research on wind pressures on high bridges, and can be seen as a very apt expression of the idealistic approach to engineering in France.

Educational Institutions, the Telegraph, and Agriculture

French engineering owed much of its distinctive style to the fact that there were training schools for engineers in France before 1800, as mentioned in previous chapters. These were reorganized in 1794–95, following the foundation of a central "school of public works," soon renamed the *École Polytechnique.*[9] Following a basic training there, which included practical subjects as well as mathematics, students went on to the more specialized courses at the *École des Ponts et Chaussées,* the *École du Génie,* or one of the other schools for military, naval, or mining engineers. The *École Polytechnique* soon became a model for technical education in other countries. Somewhat less ambitious polytechnics were founded in Prague (1806) and Vienna (1815), then in several major German towns from 1825 onward. In the United States, engineering education at West Point was reorganized in 1818 on the French model. In Britain, a new school of military engineering was established at Chatham in 1812, but civil engineers continued to receive very little formal training through much of the nineteenth century, acquiring their skills through what was called the "pupilage" system.[10]

The growth of institutions for technical education gave new scope for idealistic trends in technology, partly because education was always to some extent insulated from market pressures. In addition, when a technical subject began to be taught through lectures and books, its procedures had to be presented in an ordered, rational way, often with an enlarged emphasis on mathematics. This point was made in chapter 7 in connection with the teaching of kinematics at the *École Polytechnique,* but most other branches of technical knowledge went through the same process, which in some cases encouraged research into scientific aspects of the material.

One episode that illustrates other ways research could develop within educational institutions concerns electricity, which by the middle 1820s had reached a stage where it was suggesting a wide range of related ideas to many investigators. Thus Michael Faraday's work in England was paralleled in America by the experiments of Joseph Henry of Albany, begun in 1826. Henry was a little ahead of Faraday in demonstrating the principle of the electric motor, and his work with electromagnets led him to devise instruments that could be used to detect signals sent over metal conducting wires. In 1831 he demonstrated an electric telegraph of this kind over a mile of wire.

It was at this stage that research within universities, still very informally organized, began to play a part. The telegraph needed improved batteries to be successful, and one possible type was devised by J. F. Daniell, the professor of chemistry at King's College, London. At the same time, one of Daniell's colleagues, Professor Charles Wheatstone, was involved in the development of a new electric telegraph, patented with William Cooke in 1837.

However, the form of telegraph that came to be most widely used was that invented by Samuel Morse between 1835 and 1844. Morse was a teacher of painting and sculpture at New

York University, and soon found he needed the help of a colleague with some knowledge of chemistry and batteries. This was provided by a professor named Leonard Gale. At the same time, a former student of the university, Alfred Vail, attended to mechanical aspects of the innovation.[11]

The most famous school of chemistry in the world at this time was at the small German university at Giessen, in the Grand Duchy of Hesse. It had been founded by Justus von Liebig, who had studied in Paris during the 1820s under J. L. Gay Lussac, pioneer of a technique for analyzing organic compounds. Use of the technique had shown, for example, that sugar, starch, and much of the material in wood belonged to a family of compounds with similar compositions, and Gay Lussac and his co-workers called these "carbohydrates."

As Liebig built up his school of chemistry in Giessen, he made a point of using laboratory work as a means of teaching. Thus Liebig insisted that students should undertake a series of analyses as part of their training, using methods developed from Gay Lussac's. The school became famous, and students came from France as well as Germany, then after 1835 in considerable numbers from Britain, and finally in the 1840s from the United States.

In both Britain and America, the teaching of chemistry underwent great changes during the 1840s. This is usually attributed to Liebig's influence, which was spread by his former students, and by a major book, *Organic Chemistry in its Application to Agriculture and Physiology.* The importance of the latter lay partly in its stress on soil minerals. Chemists had previously emphasized organic components of the soil as the source of fertility. But in the eastern United States, where farmers were facing severe problems of declining fertility, they knew from experience that some mineral fertilizers such as gypsum could be helpful. When Liebig's book appeared in an American edition in 1841, it was immediately welcomed for the light it shed on this. In the tobacco lands of Virginia, its central point was quickly applied to show that declining soil fertility was due to exhaustion of alkali minerals such as potash, lime, and magnesia. The book quickly acquired an enormous reputation, and as one historian has put it, "agricultural chemistry" became a symbol of the "usefulness of science."[12]

Liebig's view of soil fertility encouraged a somewhat excessive enthusiasm for chemical analysis of soils, with many tests yielding almost meaningless results. However, a corrective came in the work of Samuel W. Johnson, professor of chemistry at Yale, who wrote two important books in the 1860s that offered a more mature understanding of plant nutrients, including nitrates. Johnson's awareness of the complexity of plant growth made him far more cautious than Liebig had been about drawing general conclusions from laboratory results. It seemed important to him that theories should be tested on special farms where controlled experiments could be done. He had seen such a farm in Germany and knew of the one at Rothamstead in England where the researches of J. B. Lawes[13] had led to the invention of superphosphate fertilizer in 1839.

Johnson was not the only person to think like this. State geological survey staff and members of agricultural societies could also see the need for agricultural experiment stations. Their arguments had such effect that by the end of the 1870s, Johnson was installed as director of an institution of this sort. Then, by the end of the 1880s, there were forty experiment stations in the United States, some associated with the land-grant colleges set up under the Morrill Act of 1862. The subsequent development of agricultural research in the United States was unrivaled in any country.

Industrial Organic Chemistry

In Britain during the 1830s there was a growing awareness of the importance of chemistry in industry as well as in agriculture. One response was the foundation in 1845 of the Royal College of Chemistry in London, a body later absorbed by other institutions, ultimately to form part of Imperial College within London University. Though financed by the state after 1853, the Royal College was initially a private venture whose subscribers included apothecaries, druggists, manufacturing chemists, and landowners interested in agricultural chemistry. Liebig's advice was sought and his assistant, A. W. Hofmann, was appointed the first professor. As in Liebig's school at Giessen, training was based on practical work in the laboratory, which combined research with education.

One challenge facing chemists at this time was associated with the rapid expansion of gas lighting that had taken place since the first successful installations had been built in Birmingham and Manchester soon after 1800. When coal was heated in retorts to produce gas, large quantities of tar were produced as a by-product. It was found that distillation of the tar yielded a range of different oils, one a useful solvent for rubber and others of value as fuel for oil lamps and preservatives for timber. To cater for these uses, a small tar-distilling industry had evolved.

Much of the work done by Hofmann and his students at the Royal College of Chemistry involved further research on the constituents of coal tar. For example, C. B. Mansfield prepared pure samples of benzene from tar distillates. His work prompted Read Holliday, owner of a chemical works in northern England, to design improved lamps for use with the coal-tar oils his firm produced. Then in 1856 another of Hofmann's students, W. H. Perkin, experimenting with toluidine and aniline made from coal-tar products, accidentally discovered an attractively colored mauve substance that he found could be used as a dye. With help from his father and brother, Perkin soon had a factory in operation to manufacture the dye.[14]

Soon other research on the reactions of aniline yielded more useful dyes. In 1859, a magenta was discovered and put on the market by a French firm, Renard frères. In Germany, four aniline dyeworks were set up within five years of Perkin's discovery, and in 1868 two researchers in Berlin, Liebermann and Graebe, discovered a way of using anthracene from coal tar to make the red dye alizarin, previously obtainable only from the madder plant. Synthetic alizarin was soon being made by the firm known as BASF in Mannheim and by the Hoechst company near Frankfurt.

The importance of these advances in chemistry is partly that their further development was associated with the emergence of a new type of institution, the industrial research laboratory. Such laboratories, employing large numbers of chemists by the 1880s, were established by a few German companies, including BASF and Hoechst, around 1870, and later by most dyeworks and some other companies. These laboratories might be said to have served German industry in rather the same way as experiment stations during the same period served American agriculture.

However, industrial laboratories run by chemical companies need to be distinguished from purely personal laboratories working on industrial problems such as the one Perkin established in his factory during the 1850s, or Edison's at Menlo Park, near New York, set up in 1876. They must also be distinguished from analytical laboratories in industry that did routine tests but no research. Steelworks sometimes had such laboratories, for example, for checking the composition of ores. One very early, short-lived example, set up in 1864, was in a steelworks at Wyandotte, Michigan; it was designed by William F. Durfee, a man who had studied science at Harvard and who saw the laboratory as "institutionalizing his own intellectual methods."[15]

In America and Britain, industrial research (as opposed to analytical work) was slow to emerge. Edison's laboratory was perhaps significant as an example, and there were also manufacturers of electrical equipment who gave the occasional inventive individual freedom to pursue his own ideas. One thinks of Nikola Tesla joining the Westinghouse company at Pittsburgh in 1886 and working on the theory of alternating currents as well as designing transformers and motors. In the Bell company, an industrial laboratory was established for testing telephone equipment, and in the 1890s began to undertake some basic research.[16]

In Britain, as the teaching of chemistry expanded in the universities and at the Royal College, attitudes were evident that must have discouraged any thought of industrial research. On the one hand, industrialists tended to distrust academic science, while on the other, chemistry professors such as Henry Roscoe in Manchester increasingly talked about "pure science"[17] as their ideal, meaning a search for knowledge unrestrained by practical demands. One might put this in perspective by remembering that engineers sometimes seemed to disregard economics in designing ideal railroads or Eiffel Towers. However, they could hardly admit openly their wish to pursue the ideal rather than the practical, nor could they claim any virtues for "pure technology." But scientists working in universities could justify "pure science" as an intellectual discipline that contributed to "liberal education."

One of the major themes in "pure" chemistry research at this time was a new conception of the bonds linking atoms together within molecules. In 1858, the idea that the carbon atom always had four bonds but the hydrogen atom only one was put forward by August Kekulé, a former student of Liebig who had worked briefly in London before becoming a professor in Ghent and then in Bonn. Kekulé examined a number of compounds whose molecules seemed to contain very stable groups of six carbon atoms and put forward the view that these six atoms formed a ring. The simplest substance to have ring-shaped molecules was benzene, and he represented the "benzene ring" both by a model and by diagrams in which atoms were circles linked by bonds shown as lines.[18]

In Germany, several discoveries of industrial importance were made possible by this new ability to visualize molecular structures. For example, Emil Fischer worked out structures for several important dyes. It then became possible to study these substances in a more systematic fashion, making known modifications to their molecules and observing the effect on color and fastness. Several new dyes emerged from this research, while Fischer went on to investigate the molecular structures of carbohydrates (from 1884) and protein (in the 1890s).

Another consequence of studies of molecular structure was the discovery of very long, chainlike molecules in physically strong materials such as cotton fibers and also in such unusual substances as rubber and silk. Cellulose was the material of most importance in cotton and timber, and experimental work had already led to the discovery of a new transparent material, celluloid, first made by John and Isaiah Hyatt in America in 1869. Further research produced a new explosive (cellulose nitrate or "gun cotton") and a cheap substitute for silk (viscose rayon, 1892).

Applied Science and Commercial Values

The relationship between science and technology defies easy definition partly because it is highly sensitive to differences in institutions and values and thus varies from one country to another. In the United States, James Renwick, a professor at Columbia College, attempted to bring science and engineering into a closer relationship through two books on mechanics published in 1832 and 1842. However, the verdict of a modern commentator[19] is that the books only served to demonstrate how scientists and engineers used mathematics and theoretical ideas with quite different ends in view. Nonetheless, trends favoring technologically oriented "applied science" can be identified in America at least from the time when the Franklin Institute began a study of boiler explosions on riverboats in 1830. Then later, the remarkable growth of agricultural research mentioned previously indicates an interest in applied science that would be hard to match elsewhere.

In Britain, attitudes differed sharply between the two nations of England and Scotland. While English chemists developed the ideal of pure science and sometimes expressed disdain for its industrial applications, Scottish culture espoused values that were altogether more positive with regard to commerce and industry. Calvinism had put down deeper roots in Scotland than England, and there was an inheritance from what chapter 5 characterized as the later form of the Protestant ethic, with its positive view of individual enterprise. Connected with this were distinctive attitudes toward education and a continuing respect for Bacon's ideas.[20]

One striking illustration of the difference between Scottish and English values has been provided by a study of university students during the period 1800–1850. While about 50 percent of those attending Glasgow University came from families involved in commerce and industry, the corresponding figure for Cambridge was only 6 percent, many of the other Cambridge students being sons of clergy or landed gentry.[21] Scotland, and especially Glasgow, was thus fertile ground for the development of applied science, and indeed there was a history of regular contact between academic science and engineering or industry going back well before 1800.

In the middle decade of the nineteenth century, Glasgow and to a lesser extent Edinburgh universities were again at the forefront of applied science after a period of reform and regeneration. It is significant that three of the men leading this movement were also involved in the most challenging technological enterprise of the period. This was the attempt to lay a submarine telegraph cable under the Atlantic, linking North America with Ireland (and hence with Britain). But another aspect of applied science at this time was the appointment of the first professors of engineering in Britain, first at Glasgow in 1840 and then in London a year later. This did not mean an immediate improvement in engineering training, because nobody in Britain had much idea as to how this should be done. It was only with the appointment of W. J. M. Rankine as professor at Glasgow in 1855 that effective university education in engineering began there.[22] With experience of railroad engineering combined with knowledge of physics, Rankine was in a good position to draw together the different aspects of applied science. In a series of four "manuals" printed between 1858 and 1869, he established an approach to the teaching of his subject that was widely influential.

Rankine's predecessor, Lewis Gordon, was not so effective as a university teacher but made his contribution as an engineer, notably when working on initial stages of the Atlantic telegraph project. Recognizing the unprecedented problems involved and the lack of expertise among those responsible, he consulted another Glasgow professor, William Thomson, who had devised a theory for the transmission of telegraph signals using equations developed by the great French mathematician Fourier.

When the first cable was completed in August 1858, its performance was very disappointing. Signals received from North America were very weak and would hardly have been decipherable but for a sensitive reflecting galvanometer devised by Thomson. Within three months the cable failed completely. Investigations demonstrated faults in the manufacture of the copper conductor and its insulation, which was made from gutta-percha, a resin-like substance obtained from a tropical tree. A. W. Hofmann, at the Royal College of Chemistry, had shown that this was a hydrocarbon, and now further analysis at King's College, London, indicated that it sometimes deteriorated by oxidation. The problem was that the laying of undersea cables had been pushed forward with idealistic fervor to fulfill visionary schemes for worldwide communications, and had outpaced technological capability. Thus when the British parliament appointed a committee to investigate the subject in 1861, their hearings showed that of 11,364 miles (18,290 kilometers) of submarine cable that had then been laid, mainly in the Mediterranean, the Red Sea, and the Atlantic, only "a little over 3,000 miles" (1,900 km) were working.[23]

In 1865 a renewed effort to establish an Atlantic telegraph was begun using Brunel's gigantic *Great Eastern* as the cable-laying ship. There were setbacks when the cable broke, but in 1866 *Great Eastern* successfully completed two cables.

William Thomson was the professor of natural philosophy (or physics) at Glasgow, and had studied at Cambridge. But coming from a Scottish (and Irish) background, he reacted against the "pure science" approach at Cambridge and took a thoroughly commercial view of himself as a "professional man possessed of marketable knowledge."[24] With two colleagues who had also devised telegraph equipment, he patented his inventions and gained financially on a considerable scale when the Atlantic telegraph was finally successful. One of these colleagues was Fleeming Jenkin, who was soon to make his own contribution to the development of applied science in Scotland when he became professor of engineering at Edinburgh in 1868 and began to teach on the "Rankine system."

It needs to be stressed, though, that commercial values did not exclude idealism in this work. Speaking about the Atlantic telegraph, Thomson expressed his "sense of the grandeur of the enterprise." Fleeming Jenkin, at a time when he was working on a cable-laying project in the Mediterranean, wrote about being excited by engineering as a "bloodless, painless combat" with inanimate matter.[25]

One further point to make is that developments in electricity, chemical industry, and steel making from the 1860s onward added up to what some authors have described as a "second industrial revolution."[26] The difference between this and the first such revolution could be characterized by saying that new kinds of *institutions* had become effective in technological development, but the conventional definition is simply to say that many innovations depended on applied science. Thus it is worth noting that the difference between the success of the Atlantic cable laid in 1866 and the failure of the earlier attempt in 1858 was almost wholly due to scientific investigation.

Another aspect of electrical technology at this time was the invention of a new means of generating electric current, sometimes referred to as the "dynamo." Prior to 1870, the telegraph had been much the most important application of electricity because it used very little current and could be run from batteries. The mechanical generation of electricity was inefficient and expensive, partly because most generators depended on permanent magnets. The new dynamo was a "self-exciting" device in which magnets were replaced by coils of wire producing the necessary magnetic field. The idea of how to do this had come to several people almost simultaneously, including C. F. Varley, who had worked with Thomson and Jenkin on the Atlantic telegraph, and Charles Wheatstone, another telegraph pioneer. The third person in this story was Werner Siemens, the leading German telegraph engineer, and his version of the dynamo was the most important because his company, Siemens and Halske, began to manufacture dynamos in Berlin around 1870.[27]

The cheap generation of electric current that was made possible by this invention opened the way to a much wider use of electricity for lighting, in industry, and in transport. Thus in 1879 Edison in America and Swan in England independently produced the first successful electric light bulbs with incandescent filaments. By 1882 Edison had small power stations operating in both New York and London, and a rapid expansion of electric lighting followed.

Coal in the Second Industrial Revolution

The new industries associated with the second industrial revolution depended on a rather narrow resource base. Coal was the most common source of energy for industry and for generating electricity, with steam engines driving dynamos (except in some cases in which water power was used). Organic chemicals were mostly derived from coal tar. With regard to the iron and steel industries, one of the biggest users of coal, major new processes for bulk steel production had been introduced in Britain in 1856 (by Henry Bessemer, anticipated in America by William Kelly), and in 1867 by William Siemens (brother of Werner Siemens, the electrical engineer). Phosphorus impurities in iron ore caused problems for both processes. However, in the United States, the availability of low-phosphorus ores from near Lake Superior made possible the growth of a very large steel industry around Pittsburgh, where the efficiency of the Bessemer process was considerably improved through the inventiveness of Alexander Holley.[28]

In Britain, the phosphorus problem was solved in 1879 by two cousins named Gilchrist and Gilchrist-Thomas, one of whom was a professional analytical chemist at the Blaenavon Ironworks in Monmouthshire. Between them, they showed that if linings for furnaces or Bessemer converters could be made of a basic (i.e., alkaline) material such as dolomite, the phosphorus impurities would be absorbed. This is sometimes regarded as the first major science-based invention in steel making, and thus as another milestone for the second industrial revolution.

Commentators who noted the dependence of key industries on coal tended to focus attention on the steam engine rather than the use of coal to make coke for steel making. For example, steam engines were used in steel works to pump the air blast to furnaces and for rolling steel. However, steam was also the major source of power in factories and for transport by ship or railroad.

In this context, the possibilities for improving the fuel economy of steam engines became a challenge to engineers and also a topic of great theoretical interest to scientists. Prominent among the latter was the applied science group at Glasgow University that included William Thomson and W. J. M. Rankine. To make progress with a theory of engines, these men had to clarify their ideas about the role of temperature (on which the Frenchman Sadi Carnot had contributed important ideas). They had also to understand how heat could be converted into mechanical energy (where the experiments of J. P. Joule in Manchester were crucial). While Rankine suggested useful ideas about the "internal energy" of gases, it was William Thomson who achieved a synthesis in 1851, thereby defining the basic principles of the new science of thermodynamics.[29]

One remarkable aspect of this work is that although thermodynamics originated in studies of steam engines (and also hydraulic machines), it yielded results of wide significance, and its laws were perceived as applying to the whole universe. The theory showed that some loss in "available energy" was inevitable in the working of engines, and this set Thomson thinking about other sources of power used by man, such as windmills, waterwheels, and fodder eaten by horses. He recognized that all these sources of energy ultimately derived from the sun, and speculated how the sun produced its heat and light. It seemed clear that there must be limits to this process, and that ultimately the sun would cool. Thus he became committed to a cosmology that "set limits to human habitation of the earth."[30]

It is a striking paradox of nineteenth-century Britain that some of those most optimistic about technology and industrial progress were also the most concerned about limits to such progress set by nature. A text for optimistic idealism, quoted in the previous chapter, had been provided by Charles Babbage in the 1830s. Though he held a university post at Cambridge, Babbage had a nearly Scottish appreciation of knowledge as a form of capital that could be invested to produce wealth. But he referred to doubts as to whether "the weak arm of man" had access to the physical resources necessary "to render that knowledge available." His answer was that there would be a succession of discoveries about new sources of energy, or ways of using the tides, or volcanic springs. Knowledge was a resource "which ages of labour and research" could "never exhaust."[31]

In discussing his own optimistic industrial vision, William Thomson echoed Babbage's views about how the "weak arm of man" was strengthened by knowledge. But in keeping with his belief that there were limits to the energy sources that sustained civilization, he thought that reliance on the steam engine would be a temporary phenomenon. In 1881 he commented that as the "subterranean coal stores of the world are becoming exhausted surely, and the price of coal is upward bound . . . windmills or wind motors of some form will again be in the ascendent."[32]

Thomson had also spoken about some of these questions at a meeting of the British Association at Manchester in 1861, and it was a future professor of economics at Manchester, W. Stanely Jevons, who soon after developed the most pessimistic analysis of all. He pointed out that during 1861, nearly 90 million tons of coal had been extracted from British mines, mostly for consumption within Britain. Moreover, coal production had been increasing at an annual rate of 3.5 percent for three or four decades. If this rate of increase were sustained for another hundred years, production would reach 2,600 million tons a year in 1961, and the total quantity of coal mined would amount to 102,000 million tons. This was clearly impossible. Britain's total reserves of accessible coal were then estimated to be less than this. What would happen, said Jevons, was that coal would have to be obtained from progressively deeper and more difficult seams, and so would become increasingly expensive. Consumption would no longer increase by 3.5 percent annually. It would soon become static, and then begin to fall. Thus, "our present happy, progressive condition is a thing of limited duration."[33]

In 1901, when Jevons's analysis was being questioned by an official inquiry into Britain's coal reserves, a new edition of his book was brought out by a friend, A. W. Flux. This showed that the rate of increase in coal production had decreased from 3.5 percent to just over 2 percent each year. Flux very reasonably argued that this vindicated Jevons's view that the higher rate could not be kept up for long. Then, in order to demonstrate that even the 2 percent growth rate could not be sustained, Flux followed Jevons's example and plotted a graph to show the implication of annual 2 percent increases. If this graph is superimposed on one showing the actual production of coal in Britain since Flux did this work (figure 40), it can be seen that output figures only follow the 2 percent growth curve until 1913, after which coal mining began a long period of decline, beginning with the cessation of British exports of the fuel during the First World War.

Jevons's view of the future of technology was a pessimistic one—too pessimistic, as it turned out—yet he combined his gloomy predictions with a resolute and almost irrational hopefulness. Coal resources would ultimately be exhausted and industrial progress threatened, but if, in the meantime, "we lavishly and boldly push forward in the creation and distribution of our riches, it is hard to over-estimate the pitch of beneficient influence to which we might attain in the present. *But the maintenance of such a position is physically impossible. We have to make the momentous choice between brief greatness and longer continued mediocrity.*"[34]

Engines and Aviation

If Jevons had been writing in America, he would probably have been one of those who were concerned that forests that had previously covered so much of the eastern part of the continent were being destroyed by wasteful forms of development. He would have mentioned other sources of energy also. In Britain, water power had hardly developed since the early part of the century, but in New England, it provided more power for factories than did steam engines, even after 1850. Moreover, the potential of water power had been enhanced by the introduction of turbines. These were a French invention, but innovations at Lowell, Massachusetts, resulted in a new type, the Francis turbine,[35] in the 1850s.

From 1859, yet another source of energy was represented by the first oil well at Titusville, Pennsylvania. The crude petroleum it yielded was distilled to obtain kerosene as fuel for oil lamps. A little later, Russia began to exploit the Baku oilfield on the Caspian Sea, again producing kerosene as a lamp oil and burning the heavy residues from distillation as fuel for steam engines. In all early oil-distilling industries, gasoline was a by-product for which there was very little use.

The first successful internal combustion engines, made in Germany from 1864, ran on gas produced from coal and were

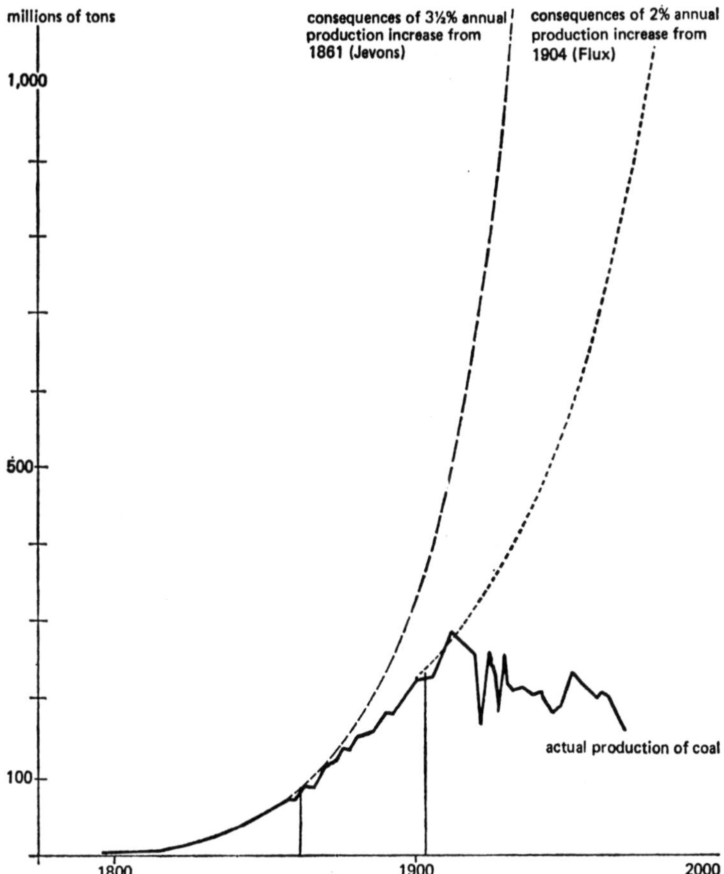

Figure 40
Graphs plotted by Jevons and Flux to illustrate the implications of sustained growth in coal consumption, superimposed on a graph showing actual consumption of coal in Britain.

(Compiled by the author from the frontispiece to W. S. Jevons, *The Coal Question*, 3rd edition, ed. A. W. Flux, London: Macmillan, 1906, and incorporating more recent statistics.)

used mainly in workshops and factories. In the 1870s, several engines were built to run on kerosene, but it was not until about 1880 that an engine was designed that could satisfactorily run on vaporized gasoline. In 1886 one was mounted on a four-wheeled chassis by Gottlieb Daimler to create the first automobile. Quite independently, but also in Germany, Carl Benz built a three-wheeled car, while in the United States, Charles E. Duryea began work on a vehicle driven by what he called "the gasoline vapour explosion engine." This led to the first American "horseless carriage" in 1891.

Another development in the 1880s was the introduction of a "safety bicycle," which unlike most earlier cycles could be ridden by almost anyone. With its chain-driven back wheel and tubular steel frame, it originated in England about 1885. Many people felt an immense sense of liberation in their first experience of cycling, especially women, for whom it offered a break with the restrictive conventions of the time. A remarkable boom in bicycle sales in the United States peaked in 1896–97, when 300 companies made over a million bicycles.

Such enormous popularity was short-lived, but bicycles helped arouse greater interest in automobiles by encouraging a taste for independent travel and stimulating the development of related inventions such as pneumatic tires and ball bearings. With the collapse of the bicycle boom in 1896, some of the workshops that had been making cycles turned to building automobiles. There was vigorous experimentation with a variety of power sources, and many electric cars (with batteries) were made, and oil-fired *steam* vehicles also. In 1900 more steam cars were sold in America than any other type, but by 1905 the gasoline-engined automobile was most strongly in demand.[36] By then, too, a new use for this type of engine had emerged when Orville and Wilbur Wright's pioneer aircraft made its first flight at Kitty Hawk, North Carolina, in 1903. This also had roots in the bicycle boom, because the Wright brothers' cycle business had provided workshop facilities and profits to pay for experiments.

Because of the close relationship between bicycles, automobiles, and early aircraft, their development can be regarded as a "movement" in technology in the same way as interlocking groups of new techniques described in chapter 7. One characteristic of this movement is that its focus was at first rather far removed from the traditional areas of industrial innovation.

Bicycles and automobiles were initially used mainly for recreation and sport. They represented a new concept of personal mobility, and at first the aircraft seemed to fit the same pattern.

Thus to understand the impulses that lay behind these inventions, we need to look beyond questions of utility and profit to consider different satisfactions, such as the enjoyment of mobility and the exhilaration of speed and power. Other inventions made around the turn of the century were also related to personal enjoyment, such as the gramophone, evolved from Edison's 1877 phonograph by 1890, and cinematography, developed in 1896. But more was involved in this movement than just the invention of new forms of recreation. Many inventors, particularly of aircraft, were primarily engaged in exploring challenging new technological possibilities. This was an idealistic motive, comparable to that of "pure" scientists exploring the workings of nature. Such activities are not undertaken for financial gain but because of the "psychic rewards"[37] of the exploration, and we need to take full account of this idealistic, investigative spirit.

It was suggested earlier in this chapter that idealistic trends in technology were often associated with institutional arrangements that tended to limit pressures from market forces. Such arrangements included subsidies and the freedom for independent study to be found within educational institutions. In the early history of aviation, examples that fit this pattern were a subsidy of $50,000 given to Samuel P. Langley by the United States War Department in 1898 for an experimental aircraft,[38] and a glider built in 1895 by a Glasgow University lecturer in engineering, Percy Pilcher. But these instances are atypical. In Europe, pioneers of aviation (and automobiles) were more often men with leisure and enough personal wealth to pay for experiments, such as Sir George Cayley in England and Otto Lilenthal in Germany. For them, flight was a matter of personal research unrelated to financial gain.

In the United States, by contrast, the institution of overwhelming importance for experiment with automobiles as well as the Wright aircraft was quite clearly the engineering workshop—especially the bicycle workshop—and the culture of skill and creativity it fostered. There were two distinct traditions, with workshop practice in the eastern states dominated from early on by the effort to achieve interchangeable-parts manufacture, first in handguns and then in sewing machines. Many gun manufacturers turned to cycle building in the 1880s and 1890s, including the Remington, Colt, and Winchester armories. A distinguished historian of these developments, David A. Hounshell, also notes that Sharp's Rifle Manufacturing Company at Hartford, Connecticut, had made not only guns, sewing machines, and bicycles, but then built automobiles for the bicycle entrepreneur Albert A. Pope.[39] In the midwest, a somewhat different workshop culture had evolved among carriage and wagon builders, and it was here that bicycle parts were first made by pressing or stamping sheet steel. As automobile manufacture evolved into a large-scale industry, it drew on both the interchangeable-parts techniques of the east and the midwest methods regarding press work (and also assembly).

To sum up, the argument here has been that institutions developed during the nineteenth century in ways that gave scope for engineers, scientists, and workshop technicians to develop technology in directions suggested by ideals and imagination, rather than being dictated by the marketplace (table 2).

Although, of course, the numerous bicycle workshops were market-driven enterprises, their owners and many of their technicians were also enthusiasts, willing to find spare equipment and allocate the time to develop an automobile design, or like the Wright brothers, to pursue some other investigation.

In larger companies with more specifically science-based technologies, this matter of making space for idealistic investigation had to be institutionalized in a more formal way by setting up research laboratories. The pioneer examples were in the German chemical industry, but the pattern was reinterpreted in new ways when the General Electric Company set up a research laboratory at Schenectady, New York, in 1900. The point about avoiding subservience to market forces was made clear when it was explicitly stated that the research done there was separate "from all (industrial) operations and commercial pressures." The first director, Willis Whitney, had studied at Leipzig, where he had experienced a scientific community "that set store by the act of knowing simply of itself."[40] Recreating this atmosphere in work on electric lamp filaments, and slowly building up a team of research chemists and technicians, he contributed not only to a practical innovation—the tungsten filament lamp of 1909, with an improved version by 1913—but his researchers also produced a great deal of new scientific knowledge of metallurgy, electric discharges in gases, and the behavior of electrons.

Table 2
Some nineteenth-century institutions with capacity for "idealistic" innovation.

Type of institution	National background	Examples of "idealistic" innovations or projects
Private companies:		
a) railroad companies	Britain and elsewhere	cathedral-like train stations[a]
b) workshop culture	U.S.A.	mechanization taken to the[b] limit (e.g., Colt, chapter 8)
.. .. ..	U.S.A.	bicycles, automobiles, aircraft[c]
Professional engineers in private practice	England	Brunel's ideal railroads[ac]
	France	Eiffel Tower[a]
Government projects and public-sector engineers	England	pumping engines as monuments[ac]
	France	rational planning of railroads[b]
Educational institutions	France	mathematical approach to engineering[b]
	Germany	chemistry research[b]
	Scotland	applied science (e.g., telegraph and thermodynamics)[b]
Research institutions	Germany	industrial research laboratories (organic chemistry)[b]
	U.S.A.	agricultural experiment stations[b]
Military-industrial complex	U.S.A.	handguns with interchangeable parts[b]
	Britain	naval engineering[b]

Notes: a. Idealism related to symbol creation.
b. Technical ideals relating to rationality and the application of mathematics and science ("intellectual ideals").
c. Technical ideals relating to "aesthetics and mobility" (see Pacey, *The Culture of Technology*, pp. 82–87).

One could hardly have a clearer illustration of the pursuit of technical and intellectual ideals and of the conditions in which they most fruitfully contributed to advances in technology.

The Great War and its Prelude

One form of institutional development this chapter has not so far noted is the early growth of a "military-industrial complex" in certain countries. This term refers to a very tight relationship between a government, its armed services, and private arms suppliers, as was foreshadowed soon after 1800 in the United States by the relationship between a department of the federal government, its two armories at Springfield and Harper's Ferry, and private gun manufacturers working to government contracts. In Britain, a similar three-cornered relationship developed between government, the Royal Navy, and private arms manufacturers in the 1880s. What initiated this development was recognition that government arsenals were not able to produce guns with the range and accuracy that the German company, Krupp, had demonstrated. But an industrialist in northeast England, William Armstrong, who believed that he was making a contribution to world peace, had reequipped his works to produce weapons of the new type. It then became inevitable that the Royal Navy would buy its guns from him, though to avoid giving him all their contracts, the Admiralty encouraged another firm, Vickers, to enter the armaments market in 1888.

Because the French were building a large navy and the Germans sometimes seemed to have more advanced technology, Parliament approved higher spending on warships. Meanwhile, close links developed between technically minded naval officers and industrial managers. The Royal Navy lost its traditional conservatism and its officers asked for performance characteristics in guns, ships, and engines that firms could provide only after considerable research and development. Thus eager naval officers forced the pace of innovation with regard to advances such as quick-firing guns for use against torpedo boats (1887) and very fast destroyers, using new types of boilers and the steam turbines invented by Charles Parsons (1897).

Accuracy of gunfire became an especially challenging problem as ranges increased. The Japanese, using warships built in Britain, made advances that enabled them to destroy Russian vessels in their 1905 war at distances of 13,000 yards (12 kilometers). Research on gun aiming and range finding on moving ships occupied some of the best brains in the British navy, and to some extent in America also.[41]

Meanwhile, naval engineering came to be at the "leading edge" of British technology, with over 10 percent of the civilian male work force employed on naval projects by 1913. By this time, the Vickers and Armstrong-Whitworth companies had become large bureaucracies, sharing contracts by mutual arrangement rather than competing. Both companies exchanged patents with Krupp in Germany and the major French arms manufacturer, Schneider, creating an international arms ring. Each of these firms supplied arms to foreign states, managing export orders to provide a more even workload than was possible by reliance on the home government's orders. Between taking over two shipyards in 1884 and 1914, the Armstrong company built eighty-four warships for twelve foreign governments, sometimes introducing technical advances that British warships did not have.

With plentiful orders and no expense spared, this was another aspect of technology protected from normal market forces. It included subjects for research where, "the abstract challenge of problem-solving" was especially absorbing, and as a result, "the arms trade attracted more than its share of technically innovative minds."[42] It had become the most vigorous area of research on the British industrial scene. Within Vickers, it was said, technology had become "an end in itself."

There were parallel developments in Germany, where Admiral Tirpitz had plans to build a navy that would rival Britain's. Progress was made with this project until 1906, when the Royal Navy launched an enormous, technically advanced battleship, *HMS Dreadnaught.* When the Germans built a similar vessel, Britain responded in 1909 with a plan for eight more "dreadnaughts."

As war loomed nearer, chemical innovations once more became important. One, the work of Fritz Haber in Germany in 1913, was a method for synthesis of ammonia by subjecting a mixture of nitrogen and hydrogen to very high pressure (250 atmospheres) in the presence of a catalyst. After 1914, much of the ammonia was oxidized to make nitric acid, enabling Germany to be self-sufficient in the manufacture of explosives instead of importing nitrates in the form of Chilean saltpeter.

Britain had planned a naval blockade before 1914, believing that if war came, it could be ended relatively painlessly by cutting Germany off from imported raw materials and export markets, thereby precipitating economic collapse.[43] However, a good deal of hurried improvisation was necessary in some parts of British industry before it was possible to manage without imported chemicals from Germany. Britain and France were even using German dyes in the manufacture of army uniforms, and a government-inspired reorganization of what remained of the British dyestuffs industry was necessary in 1914–15.

In these and other ways the Great War throws an interesting light on the argument central to this chapter, that developments in chemistry research, automobiles, and aircraft often took place in an institutional context that enabled innovation to follow an exploratory, idealistic path. One effect of the war was to change this institutional context radically, so that immature technologies such as aircraft (and radio[44]) were forced to develop in different, more practical directions, and very fast. Thus in some fields, the Great War led to significant changes in the way inventions were perceived. For example, automobiles had evolved almost wholly as means of personal transport. Henry Ford had begun to make them far more widely available by means of his assembly-line manufacturing techniques, introduced in 1913. However, the potential of motor trucks for carrying freight had hardly been exploited anywhere. The development of vehicles and engines for this purpose accelerated rapidly after the war began, and by 1917 America had exported 40,000 trucks and ambulances to the Allies in Europe.[45]

As to aircraft, it is said that only 49 had been manufactured in the United States before 1914. By the end of the war, this figure had risen to 14,000 as a result of a crash program based on a standard engine.[46] In Britain, there had been rather more work done on aircraft design in the five years before war began, some of it instigated by government. But as it became obvious in 1914 that war was imminent, a hurried decision was taken to order one relatively untried type in quantity. Difficulties arose with some contractors who undertook to build the machine without any prior experience of aircraft construction, and assembly-line techniques were not used, but large numbers of planes were successfully produced.

In these and other ways, a number of prewar developments lost their idealistic or recreational character and were brought into wider use within industrial economies as well as by the military. The results included advances in air transport, radio, and motor vehicles and an extension of industrial research.

10

Idealistic Trends in Twentieth-Century Technology

Rationalized Production and Interwar Optimism

An ideal of rationalism in manufacturing whose influence has been traced in the British industrial revolution and in American work on interchangeable parts reached a point of what seemed ultimate triumph in 1913. In that year the first assembly lines in Henry Ford's factories in Detroit were installed, leading to spectacular economies in manufacture and lower prices for the already-popular "Model T" car. The term "mass production" was coined to describe the methods employed, and automobiles were thus made cheap enough for almost anybody to afford one, yet at the same time Ford workers were paid exceptionally high wages. Indeed, high wages had been found necessary to retain workers who initially reacted strongly against the stress of machine-paced, assembly-line work.[1]

Mass production became a commanding paradigm for technology during the interwar years—an example people felt compelled to respect—because it seemed to show how the *technical ideal* of a wholly rational system of machines, many of them automatic (or partly so) could also be a means of realizing a major *social ideal*: the defeat of poverty and the provision of a good life for all. Too often, it may seem from previous chapters, social ideals for technology, concerned with meeting human needs, have been in conflict with technical ideals to do with rationality and the aesthetic and other fascinations of materials and machines. One example was the dispute over Chadwick's sanitary reform program in Britain in the nineteenth century (chapter 8). With mass production and perhaps with other twentieth-century innovations, there was the appearance that these divergent trends had at last been brought together. Henry Ford certainly promoted a view of this sort in his book *My Life and*

Work (1922), and other enthusiasts proposed "Fordizing" all of American business and industry.

In the early 1920s, while hopes were still high for creative reconstruction following the Great War, architects were especially closely involved in the exploration of technological possibility, and the best of their work gave expression to the hopeful mood that then prevailed. Some of them certainly had a vision of how social and technical ideals could complement one another.

Architects were affected by the new technology of the period partly because they had new materials to work with, partly because of prospects for mass producing building components, and partly because of the implications of the automobile for the redesign of cities. With regard to materials, steel framing for buildings had developed during construction of skyscrapers in New York and Chicago from about 1890. Modern reinforced concrete technology was largely a French innovation, with many of the earliest reinforced concrete buildings in Britain as well as in France based on methods patented by Francois Hennebique in 1892–93. Concrete bridges of memorably elegant form were pioneered in Switzerland by the artist-engineer Robert Maillart, a former pupil of Hennebique.

Among architects interested in concrete, there was also a visionary response to the automobile. "Futurists" saw the possibility of less crowded cities with green spaces between tall buildings, and efficiently designed roads. Later on, in 1922, the Swiss architect Le Corbusier produced the first of several schemes for "dream cities," or "contemporary towns" in which people's rights to light and air would be respected, and where the urban environment would be free of congestion.

It was in the building of such cities that architects saw the potential for standardized, mass-produced components, designed so that large buildings could be put together quickly from factory-made parts. A limited range of prefabricated buildings using cast-iron columns and beams and corrugated sheeting for roofs had been produced during the nineteenth century, but now the possibility was seen for a far more extensive use of factory methods. Le Corbusier wrote about "mass-production houses" in his influential book, *Vers une architecture,* published in 1923, which proposed a new architecture for the machine age.[2]

The fact that mass production of household objects and whole systems of building components would mean little scope for decoration was accepted as a challenge by many designers. They talked about a "machine aesthetic," and an international exhibition of decorative arts held in Paris in 1925 promoted the idea. But even before 1914, the German architect Walter Gropius had discussed standardization in design to make possible the industrial production of small houses. In 1919 he opened a reorganized art school at Weimar, which became famous as the Bauhaus. It was to be a combined school and workshop, in which principles for design in the machine age were to be worked out.

The Depression of the 1930s introduced a new social concern into these discussions. With many people unemployed and sinking into poverty, the challenge was to reduce the *cost* of housing by prefabrication. In America, Foster Gunnison was the most interesting of experimenters in this field, first with a steel-framed, asbestos-cement house that was produced between 1932 and 1935, but not in sufficient numbers to justify mass production.[3] Next he designed components for timber houses that could be erected in different ways to produce dwellings of a dozen different kinds. Some 4,500 houses had been built on this system by 1941.

In Britain, similarly, the 1930s resulted in deprivation and malnutrition on a scale that shocked many people, and a few of the best new buildings of the period were symbolic of this concern. In London, the Peckham Health Centre was a pioneer example of the new architecture, and at the same time a model for the development of health care. In rural Cambridgeshire, a new concept of "village colleges" combined schooling for children from groups of villages with workshops for technical training and facilities for adult education. Walter Gropius, having left the Bauhaus following the rise of Hitler, designed buildings for one of these colleges with Maxwell Fry, a British architect. Fry saw the potential of this kind of college in strongly idealistic terms. It would make possible a new standard of education that would give people "the power to live more fully, to see their world clearly, and make decisions which are their own."[4] Meanwhile, with the onset of war that led to the destruction of many buildings by bombing, there was an opportunity to replace large areas of squalid, smoke-ridden, inadequately serviced urban sprawl and build something better.

Visionary plans for rebuilding London were produced by Maxwell Fry and his associates. They advocated replacing vast areas of decrepit brick houses by high-rise apartment blocks, generously spaced to make way for trees and grass and a more adequate highway system. The plans had close similarities with the visionary ideas of Le Corbusier and those who thought like him, and in the detailed design of standardized, factory-made kitchens and bathrooms it is clear that the machine aesthetic was to rule. But this was more than just an adaptation to London of earlier academic discussions among architects. It was also the result of the stirring of social conscience associated with reactions to the Depression.

Scientific Planning

Mass production at Ford's factories was essentially *planned* production. Every detail, including the layout of machines on the factory floor, the quantities of raw materials ordered, the dimensions of parts manufactured, and the number of components produced were all precisely and rationally planned. With economic troubles mounting during the 1920s, the Wall Street crash of 1929, and the Depression of the 1930s, Fordism seemed a failure to some people. It had contributed to the Depression by overproduction. To others, however, a more logical reaction was that the whole economy of an industrial nation needed to be planned with the same rationalism as the workings of the Ford factories.

Although not everybody went as far as that, one slice of the American economy in which comprehensive and rational planning was introduced was the Tennessee Valley, and David Hounshell, the historian of mass production, notes how the initial report of the Tennessee Valley Authority—the TVA—used the language of Fordism in arguing that by producing electricity cheaply and on a very large scale, they would contribute to raising standards of living for all.[5]

The TVA was also a significant example for a group of British scientists who saw the Depression as a tragedy that could have been avoided by rationally planning industrial development and by the wider application of science. Most of the group were either socialists or else Communist Party members, largely because they saw socialism in some form as the only political philosophy that harnessed rationalism to social and humanitarian goals. Indeed, some of the group described their creed as "scientific socialism."

One of the most persuasive authors to present this viewpoint was J. D. Bernal, who wrote an eloquent book, *The Social Function of Science,* which was published on the eve of World War II. This asserted that science "is the chief agent of change in society" and claimed that in a rational and planned society, science would be able "to solve completely the material problems of human existence."[6] The argument was conducted in terms of "science" rather than "technology" because of an assumption that science was the chief source of technological innovation. Thus, where people were unemployed or starving, this was because of a failure to use science properly. The small amount spent on science in Britain illustrated the problem. Its allocation was biased against biology, Bernal said, and what discoveries were made in chemistry or physics were left unapplied. A specific example he quoted from outside Britain was the effect of the synthesis of indigo dye and its industrial production at the end of the 1890s. Growers of indigo in India were ruined. "It is said that one million Hindu . . . labourers died of starvation." The moral of this for Bernal was that the application of science in the *absence* of planning, in this case for alternative employment, would often lead to chaos.[7]

The outbreak of war gave Bernal and his colleagues a fresh determination to see science properly used, a view expressed in a joint but anonymous publication they wrote with great speed in 1940 entitled *Science and War.* Then while some of Bernal's colleagues contributed actively to the development of radar, one of the most important technological innovations to emerge from the British war effort, Bernal himself, together with the biologist Solly Zuckerman, was taken on by Lord Mountbatten, Chief of Combined Operations, to make a study of the effectiveness of bombing. This proved to be an extremely fruitful assignment during which Bernal "was able to work out in practice what he had up to then considered only in theory."[8] The outcome was not only better planning of some bombing raids and a contribution to the design of Mulberry Harbour, but also the development of what became known, from 1941, as "operations research." This is one of the tools of modern industrial management (and a form of planning), and its founders are regarded as including Bernal, Zuckerman, and C. H. Waddington.[9]

When the war was over, Bernal was given the task of planning conversion of a factory that had produced tanks so that it could now manufacture prefabricated houses. "We shall plan to do with housing exactly what we did with weapons of war," he said, indicating the urgency with which he intended to start production.[10] Large numbers of single-story houses made of asbestos cement and metal were produced in several centers. They were intended only as temporary replacements for houses destroyed in the war, but in their modest way were a highly successful example of the industrial production of buildings. They offered good standards of hygiene, warmth, and comfort by comparison with much other housing at this time, and were well liked by those who lived in them.

Science in World War II

Bernal and the colleagues who shared his social philosophy were nearly all involved in university science in Cambridge and London.[11] Bernal's field was X-ray crystallography, a technique that was enabling more precise ideas to be formed about the structures of complex molecules. For example, biochemistry research at Cambridge during the 1920s and 1930s had led to important discoveries about vitamins. Some of the X-ray studies carried out by Bernal and his colleagues revealed the complex molecular architecture of these substances. Much later, other colleagues contributed to discovery of the structure of the DNA molecule[12] whose fundamental role in genetics had recently been recognized.

In the 1930s, Bernal's equipment was located in the Cavendish Laboratory in Cambridge, which was one of some half-dozen centers in Europe—including also Göttingen, Berlin, and Paris—where exceptionally fruitful research had revealed much new detail concerning the basic atoms out of which all matter is built. In 1932 one of the Cambridge workers discovered the neutron, an important component of the atomic nucleus, and then in 1938 Otto Hahn and colleagues in Berlin showed how a uranium nucleus could be split by the impact of a neutron. It is remarkable that Bernal and his friends, as practicing scientists in this environment, were able to develop their social philosophy concerning the uses of science and to write about it eloquently and well. Many of their colleagues were so absorbed in demanding and excitingly creative research that they had little time to think about anything else.

For many people, this was a golden age for science. But the spell was broken by Hitler's rise to power in Germany and his vicious campaigns against Jews. The heart went out of research at Göttingen University when seven scientists were forced to leave in 1933 because of their Jewish descent, and the same thing was repeated elsewhere. Einstein had left Berlin in 1933 for Princeton in the United States, and many other displaced scientists followed him to America. Germany had already lost its early preeminence in industrial chemistry to the large and active laboratories of American firms (in one of which nylon was first made in 1935). Now Hitler's policies were ensuring that America would become preeminent in nuclear physics also.

Because nuclear fission had been first recognized in Berlin in 1938, there seemed a real danger that Hitler's government would learn of the potential for a fission bomb and would develop such a weapon. The story of how the United States government was alerted to this danger by a letter signed by Einstein is well known. Soon the British government was given more specific details by two refugee scientists based at Birmingham University, Peierls and Frisch. They had calculated that the amount of uranium-235 needed to make a bomb was much smaller than previously thought. This meant that a nuclear weapon now seemed a much more practical possibility. Serious research on the production of a bomb began in Britain in mid-1940 and in the United States just over a year later.[13]

At first the work was done on university campuses, though with unprecedented government funding, but when the American project moved from the research to the production stage in mid-1942, large new facilities were required, and the whole scheme was put under military control, with Colonel (later General) Leslie R. Groves in command. This was a disturbing experience for the scientists who were now subject to strict rules of discipline and secrecy that were quite foreign to their normal way of working. The establishment where the first bombs were designed and put together, however, at Los Alamos, New Mexico, was more of a research laboratory, and was run by a research scientist, Robert Oppenheimer. It was in an isolated place, surrounded by strict security, but within it Oppenheimer maintained the spirit of free discussion characteristic of the pursuit of pure science in a university.[14] These arrangements insulated the scientists from outside pressures just as effectively as university research, and though the consciences of some were disturbed by the awesome implications of the weapons they were designing, conditions were in most respects good for the scientists to become totally absorbed in their work.[15]

No nuclear bomb was being developed in Germany, but a somewhat similar project was at work on military rockets at a research station at Peenemünde on the Baltic coast. In this case, many of the scientists and engineers involved, led by Werner von Braun, had a long-standing interest in the potential of rockets for space exploration and had in some respects welcomed the militarization of rocket research as a means of gaining more resources for a pet project.[16]

The significance of these "mission-oriented" projects needs to be placed in a broader context than the military imperatives of World War II. Peenemünde and Los Alamos should be examined in relation to other institutional arrangements (such as those discussed in the previous chapter) that have had the effect of partially freeing groups of scientists and technologists from economic and social constraints. Contrary to the view that invention is an economic activity, creativity is often best encouraged by an atmosphere free of economic pressure and by maximizing the "psychic rewards" of the work. In a whole range of institutions from Liebig's laboratory in nineteenth-century Germany (chapter 9), through development of the tungsten lamp filament at General Electric,[17] via Los Alamos to the modern computer firm's laboratories, one finds the same pattern. Creativity in science and technology is related to mutual stimulus among colleagues in work that is highly focused and "technically sweet." That means it is usually highly idealistic work—in the sense of "technical ideals"—but at the same time is socially detached.

What, therefore, is of particular interest is to understand situations in which scientists or technologists were forced to think about the implications of their work sufficiently deeply to relate it to some sort of social philosophy. For Bernal, it would seem that ideas about the rational and scientific aspects of socialist planning created a bridge between social values and the ideals of scientific research. Waddington indicated that his politically active wife had led him to think more deeply than he would otherwise have done about social aspects of his work.[18] But it is clear that the sense of social crisis evoked by the Depression and by war was also an important catalyst.

Policies for Science after 1945

Scientists and engineers with experience of the new technologies developed in World War II reacted in very varied ways to the changed condition of the first postwar decade. For some, there was a sense of tremendous exhilaration. Accounts of what went on in the American weapons laboratories[19] and the autobiographies of some men active in the 1940s and 1950s, notably the mathematician Freeman Dyson,[20] indicate a highly stimulating atmosphere. Fundamental research in physics interacted with speculative inventions for space exploration and with practical development work on a new and even more fearsome weapon based on nuclear fusion (the "hydrogen bomb," first tested in 1951.)

By contrast, many scientists in America were appalled and saddened by the effects of their bombs when two were dropped on Japan in August 1945. As one observer put it, "their very success as scientists had made them fearful or pessimistic as citizens."[21] That is a poignant comment, indicating clearly the conflict so often seen between social ideals that citizens might support and the narrowly "technical" ideals of exploration and endeavor in science and technology.

Another, more practical postwar dilemma in America was what to do with the enormous organization for supporting scientific and technological research that had evolved during the war. Most laboratories and installations connected with nuclear technology were transferred from military control to the civilian Atomic Energy Commission. The first chairman of this body, David E. Lilienthal, had previously been head of the TVA and described this new government atomic energy monopoly as "an island of socialism in a . . . free enterprise economy."[22]

Considerable debate ensued about the role of government in other aspects of research. While it was agreed that public funding for science should not drop back to its low prewar level, there was concern about the independence of research and the dangers of political interference. One document that strongly influenced thinking on these matters had the significant title, *Science—the Endless Frontier.* Commissioned by the president, it was written in 1944–45 by Vannevar Bush, an electrical engineer and physicist who had invented circuits for automatic dial telephones, but who had also been involved in management of defense research during the war. Bush argued that the allocation and use of public funds for research should not, in peacetime, be directly controlled by government but by an independent board, and that the agenda for research should be decided by the scientists themselves. His idea was that "all who have the ability to explore" on the "frontiers of science" should have the freedom to advance as far as they could without constraints from administrators or representatives of the public (who would have little idea what promise or potential lay ahead).[23]

This language took up a theme that had been current in America since the turn of the century, when it was first argued that now geographical exploration and settlement of the North American continent was complete, science could be seen as a new frontier whose exploration would be both an idealistic quest and, like geographical exploration, a means of discovering new sources of wealth.

Although Bush in America had a far more direct influence on policy than Bernal in Britain, both had a considerable impact on debates about science and technology in their respective countries. Both expressed idealistic views, but while Bush's ideals for technology were oriented to ideals of exploration in areas outside politics, Bernal asserted a *social ideal* and wanted to see scientists and scientific method brought into government.

Another way of appreciating Bush's outlook is by referring again to the previous chapter, where we noted institutions evolving in ways that *insulated* research and technological development from too much exposure to market pressures. Now we see the idea being expressed that "science needed to be protected from politics by being insulated from executive control."[24] When the National Science Foundation began to fund research in the early 1950s, one of the chief functions of its board was to provide "insulation" between the source of funds and the scientists. All this made science and related branches of technology relatively unresponsive to social purposes and democratic debate, and elevated *technical ideals,* expressed in "frontier" imagery, to a position of primacy.

Meanwhile, atomic (or nuclear) energy posed special problems of balance between an inescapable need for some political control and a desire to explore its potential and carry out further research. An international system for sharing information but preventing construction of further weapons was drawn up by the United States government and became one of the first major issues discussed by the newly formed United Nations Organization in 1945–46. At first delegates felt that agreement would be possible, but then the Soviet Union blocked further progress. Its own nuclear project was proceeding too well for the proposed international controls to suit its purposes.[25]

One further attempt to resolve this dilemma and show how nuclear technology could contribute constructively to human development was made at the United Nations in December 1953. The American president, Dwight D. Eisenhower, suggested a new policy of international cooperation that would involve developing and sharing civil applications of nuclear energy. This was his "atoms for peace" program, and in the discussions that followed, there was much excited speculation about medical and transportation-related applications of nuclear technology. The effect of this policy was that several nations set up their own Atomic Energy Commissions. Japan signed agreements on nuclear cooperation with the United States and Britain, and Indian scientists came to America for training in nuclear physics.

The most promising civilian application of nuclear energy was for the generation of electricity, but while offering to share this apparently benign technology, the United States itself had not yet built a nuclear electricity plant. Much work had been done, however, on nuclear submarines, and it appeared that the pressurized water reactor designed for these vessels could be scaled up for electricity generation. Despite research on more promising, possibly safer reactor designs, this type of reactor was used in America's first civil nuclear power plant simply because it was in an advanced state of development and available for early application.[26] But the reactor set a precedent that was followed in most subsequent civil nuclear power projects.

Before this, Britain had secretly embarked on a program to make its own nuclear weapons, and in 1956 a nuclear reactor built to provide plutonium for the project began producing electricity for the public supply. Since the first American nuclear power station was not yet complete, this *military* installation was spuriously claimed as the world's first *civil* nuclear electricity producer. In fact, the Soviet Union had a slightly earlier plant.

Utopian Architecture: 1950–1970

In Europe at this time, attitudes to technology were mostly optimistic because of the rapid pace of postwar reconstruction, aided by American funds under the Marshall Plan. The ethos was indicated by the 1951 Festival of Britain exhibition and the very positive way it presented technology—including everything from agriculture to radar. The buildings on the festival site were also good examples of "modern" architecture. Their fittings and detail established standards of design for furniture, fabrics, and industrial products that had wide influence. Meanwhile, egalitarian ideals for housing, health, and education were expressed in the architecture of carefully planned new towns, while in London itself, some new housing took the form of high-rise apartment blocks with space for trees and roads arround them, very much in the spirit of Le Corbusier's prewar idealizations and Maxwell Fry's more recent plans. There was a clear intention among city planners that rebuilding in the capital should be mostly along these lines, and in view of what happened later it must be said that early schemes at Roehampton, Pimlico, and elsewhere in London were well built and attractive.

Other European countries, with more serious problems of postwar reconstruction, had developed industrialized building techniques to speed up rebuilding. Some ideas were derived from Le Corbusier, the Bauhaus, and a miscellany of prewar experiments, but new "systems" for building with standard prefabricated components also appeared in Denmark (the Larsen-Nielsen system), France, and other countries.[27] In Britain, however, the prefabricated houses of the 1940s were not taken further. Ordinary houses were built mainly by traditional methods and there was much dissatisfaction with the slow rate of construction.

By 1960, people were also realizing that much of British industry was no longer competitive now that many factories in continental Europe had been rebuilt. This issue and the originally separate concern about housing came together in the General Election campaign of 1964, which the Labour Party won with promises of the "white heat" of technological revolution. In some respects, Labour's emphasis on science and technology represented a revival of the approach Bernal had advocated in 1938. For example, Patrick Blackett, an associate of the Bernal group during World War II, was one of several leading scientists who were advising the Labour Party.[28]

In keeping with the approach Bernal had favored, there was initially an effort to do more economic planning in Britain than had hitherto been usual. There were also plans to extend the civil nuclear energy program, to build aluminium smelters, and to encourage the small computer industry. However, the government's strong social commitments also meant that a solution to the housing problem had to be found, and quickly. The number of new dwellings required was felt to be attainable only by "system building" and industrialization. But such was the sense of urgency that techniques were adopted for large-scale use without sufficient trial and development. The Larsen-Nielsen system had been used in Copenhagen to build excellent apartment blocks, but only to a limiting height of six storys; now it was employed for tower blocks of twenty storys and more. With some of the other building systems used (e.g., the Jespersen, figure 41), there were difficulties about quality control in the production of prefabricated wall panels.[29] As in the early days of interchangeable parts in America, it was hard to make standard parts with sufficient accuracy. When concrete panels and floor slabs failed to fit on site, incorrect and dangerous assembly procedures were sometimes improvised to keep construction to schedule.

Crisis came in 1968 when a relatively minor gas explosion in a London tower block known as Ronan Point led to the total collapse of one corner of the twenty-two-story building. Other system-built housing was by this time throwing up a range of technical, and more significantly, social problems. From this time onward, then, industrialized building and high-rise dwellings became steadily more discredited. As one commentator put it, a "huge overestimation of prefabrication" damaged the lives of countless people who were tenants in these public housing schemes. But in the 1960s, enthusiasm for industrialized building seemed "natural and proper" because of the ethos of "white-hot technology" and of a "new professional elite" that was expansionist, optimistic, and "confident in science."[30]

The genuine idealism of architects and scientists from the interwar years onward, and their visionary hope that a better social order could be built around machines, mass production, and science, had evidently lost its way. There was a cruel but instructive irony here in that the professionals betrayed their

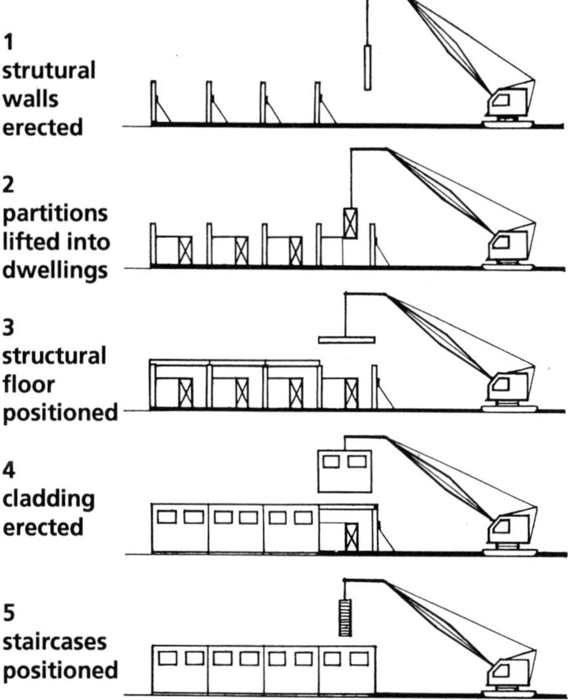

Figure 41
System building—a diagram based on the Jespersen system used in large housing projects in Southwark, London. The "system" consisted of a set of standard components such as wall panels and floor slabs. In this instance, the components were mainly of concrete and were prefabricated on an "industrialized" basis, then assembled with the help of a crane. This shows the ground floor of a building that would eventually have had five storys or more constructed with the same components.

(Source: Barry Russell, *Building Systems, Industrialization and Architecture,* London: John Wiley, 1981; reissued, London: David Fulton, 1987, Figure 8.3.31. Reproduced by permission of David Fulton (Publishers) Ltd., London.)

own good intentions. They had genuinely attempted to respond to an urgent human situation, and they could with justice argue that the speed with which they tried to achieve results was one reason why things went wrong. However, an equally significant factor was that housing needs had been interpreted only by "experts"—by an elite—and many of the social aspects of urban design had been neglected. Maxwell Fry, as prone as any other modernist architect to visionary planning, had at least proposed that "design boards" should be responsible for new housing, and should include potential occupants of the houses (especially "housewives") as well as architects and engineers.[31] Thirty years after Fry, one reaction to Britain's housing tragedy of the 1960s has been discussion of "community architecture" in which users of buildings largely control design.

The problem in postwar Britain was that many people were still influenced by ideas about scientific planning in which "experts" were always in charge. It was openly admitted by one author that a scientifically ordered society would have little room for democracy because rationality was inevitably "totalitarian."[32] Yet we now see that rationality and other "technical" ideals are often in conflict with social ideals. Democratic processes in which technologists and citizens listen to one another are perhaps the only way in which such conflicts may be resolved.

Another point to notice about the way people were thinking during the postwar period was that they tended to talk always in terms of what "science" might achieve through intelligent application. There was a failure to realize that by confusing science with technology, important processes for refinment of technique and development of hardware were being overlooked. In science, some things can be worked out from first principles, and all too frequently in the 1950s and 1960s, the British (in particular) attempted to practice technology like this. The rush to build high-rise flats that looked good in principle but that had hardly been tried in any pilot project was not the only example. The same kinds of short-cut were taken in aircraft and weapons programs, and in civil nuclear power projects, particularly with the ill-fated advanced gas-cooled reactors. The first of the latter was ordered in 1965, but neither this nor its successors ever achieved the output, reliability, or economy its designers had claimed.

It was a tragedy for Britain that so much of the good will, so many of the worthy ideals, and so much of the creative potential of the immediate postwar years was frittered away in ill-thought-out projects, leaving the country socially divided and dispirited about its future. Across the Channel, however, France was beginning to succeed with the kinds of scientific planning and technology in which Britain had signally failed. "System" building was never such a disaster there, and by the mid-1970s it was apparent that France had almost the only successful nuclear power program in the world.[33]

Runaway Technology and Alternative Ideals, 1960–1983

Throughout the postwar period, there was buoyant economic growth in most of the western world, and also in Japan. It was a boom that only ended with the oil crisis of 1973. This long period of prosperity led to wide ownership of motor vehicles and a growing range of consumer goods. Part of the background was an expanding research effort in science and technology, not only in private industry but also supported by governments. In the United States, federal spending on research and development in the second half of the 1950s was increasing at the rate of 15 percent per year, and at the start of the 1960s, under President Kennedy, the annual rate of increase rose to 16.6 percent. After 1964 there was much less expansion, but federal spending on scientific research and technology remained substantial.[34]

It was significant, however, that although a good deal of this money was spent on fundamental research in universities, the majority of the increased funding was handled by just two arms of government—the Department of Defense (DOD) and the newly formed National Aeronautics and Space Administration (NASA). This is not surprising when we recall that the late 1950s saw major developments in the arms race between the Soviet Union and the West. Both sides had built up stockpiles of nuclear weapons and large fleets of aircraft capable of delivering them.

Research was also in progress on ballistic missiles, the principle of which had been demonstrated by the V-2 rockets used by Germany in the Second World War. However, it was a considerable shock to the rest of the world when in 1957 the Soviet Union tested its first intercontinental missile capable of reaching United States territory, and used related technology to place the first artificial satellite in orbit—the "sputnik." Part of the massive increase in American government spending on research in the late 1950s was a response to these developments, aimed both at enhancing work on ballistic missiles and at promoting a more serious space program.

One other aspect of the arms race, however, was that while the superpowers competed with respect to advanced technology yet were effectively "deterred" from threatening each other too seriously by the fear of mutual destruction, small-scale wars in "Third World" nations increasingly became an outlet for superpower tensions. Some eighty relatively small local wars were fought between 1945 and 1980 and killed perhaps 8 or 10 million people, impoverishing many more.

However, the growing technological sophistication of weapons over these years was not just due to the response of governments and military planners to Soviet threats and local wars. Within the "military-industrial complex," strong relationships had developed between industries manufacturing weapons, specialist research laboratories, and the armed services. Mutual stimulus in the design of new military hardware then meant that weapons systems were sometimes pushed forward in technical capability beyond the point thought desirable by either governments or military leaders. It is sometimes argued that governments and industry give scientists their privileged, "ivory tower" laboratories in order to manipulate and use scientific activity for their own purposes. The point to notice about the military-industrial complex, however, is that interactions within it are mutual, and sometimes it has been the scientists and engineers who have manipulated the statesmen and the generals.[35]

The momentum of technological development gathered pace in the 1960s partly for this reason, partly because of the growth of consumerism, and for a variety of other reasons. To match this accerating pace of what people regarded as "technical progress," the ideology of technology had to be constantly restated, in part by the creation of suitable symbols. Thus it is noteworthy that several American presidents launched strongly idealistic technological projects in speeches remembered for their ringing and hopeful tones. Eisenhower's "atoms for peace" speech in the 1950s has already been mentioned. Two others are John Kennedy's commitment to a space program that would place a man on the moon by the end of the 1960s and Ronald Reagan's "star wars" speech in 1983.

Although motivations for the 1960s' drive to reach the moon were related to competitiveness with the Soviet Union, and much of the technology involved had military applications, the overall thrust of this project was an idealistic concern with the frontiers of adventure and exploration. Kennedy himself, in 1962, quoted the explorer George Mallory about reasons for climbing Everest: "because it's there." He also made passing comparisons with the fifteenth-century voyages of discovery when he described space as "a new ocean" on which "we must sail."[36] These voyages had opened new frontiers for Europe in the Americas and Asia, as space was now being opened up.

However, by the time this project triumphantly and thrillingly achieved its goal in 1969, there was a widespread and paradoxical sense of disillusion with much of the technology it represented. One book widely read in the 1960s (though written before 1954) argued that "technique" has superseded science, because people could no longer conceive of science without a technical outcome. "As soon as a discovery is made, a concrete application is sought," its author Jacques Ellul commented, adding that there was now a "state of mind" that scientists found hard to resist, "which makes technical application the last word."[37] Technique had become "autonomous," feeding on itself rather than responding to social priorities and human needs.

Much that had happened by the end of the 1960s made people feel that this was a correct analysis. Technological development seemed to have run away, out of control. Drugs, food additives, and agricultural chemicals were having all kinds of unintended, unforeseen, and threatening effects. There was growing controversy about nuclear power and airport noise. The Vietnam War was a source of spreading disillusion and at times seemed itself a manifestation of technological failure.

Among many and varied reactions to this mood, one of the most constructive was a new effort to foresee and manage the consequences of technological change and to build institutions and procedures for the "social control of technology."[38] In the United States, some notable milestones were the National Environmental Policy Act in 1969 and the establishment of an Office of Technology Assessment in 1972. At the same time, campaigns on specific issues led to cancellation of the American project to build a supersonic airliner (the SST) in 1971, and a tightening of safety measures at nuclear power stations that was sufficiently stringent to slow very markedly the rate at which new ones were constructed.[39]

New Social Ideals

Even more interesting, however, were some less orthodox efforts to find a new approach to technology that would make it serve social ideals (now encompassing a growing concern for the environment). One notable "father figure" to this movement was Buckminster Fuller, who had designed a system of industrial house building in 1945, but who was better known for his geodesic domes (figure 42). These manifested the technical ideal of a building structure that combined maximum strength with minimum material. The first such dome was built by Fuller's students in Chicago in 1949, and then in the next decade similar domes were used in exhibition buildings and to shelter radar installations in the Canadian arctic.

Figure 42
Impressions of a festival of alternative technology held at Bath (England) in 1974 and 1975, showing two geodesic domes, several wind turbines, and a solar energy device with a parabolic reflector.

(Author's sketch based on the "Comtek" festival, 1974, with detail added from plans for 1975.)

In the 1960s, however, the technical ideal these domes represented converged with a new social ideal. Fuller always had faith in the younger generation and took every opportunity that presented itself to lecture at colleges and universities.[40] In 1967 some who had heard him speak were inspired to build homemade versions of the geodesic dome as homes in which to practice experimental, communal lifestyles. The clearest expression of what this movement was about came with the publication of a first edition of *The Whole Earth Catalog* in 1968. The brainchild of Stewart Brand, who had studied biology at Stanford, this was in part a consumer guide to equipment likely to be needed for living outside the conventional patterns of the affluent society. In part, too, it expressed philosophical ideas about the importance of seeing the world whole, rather than breaking it down into specialized areas of technical study—and exploitation. The first page announced, "The insights of Buckminster Fuller initiated this catalog," and he was introduced as a prophet of "whole systems."

Another key figure in the effort to reestablish social ideals in technology was Fritz Schumacher, a German economist who had settled in Britain, where he contributed to planning for the postwar welfare state as well as advising on reconstruction in Germany. A spell as a consultant economist in Burma provided an important stimulus to his thinking. He became aware of different ideals for economic (and technological) development implicit in the Buddhist culture of Burma.[41] He also came to recognize that efforts to use western technology to promote economic development in non-western countries often ended in technical failure or social dislocation. He was therefore provoked into defining more satisfactory forms of technology. He wanted these to be "nonviolent" and socially benign, but the criteria he specified most precisely were related to the availability of capital and labor. In most low-income countries, he argued, the techniques that could be most constructively used would be *intermediate* in capital cost between traditional "low" technology and western "high" technology.

In 1973, Schumacher's ideas began to reach a wider audience as a result of the publication of his book *Small is Beautiful,* but before that, in 1965–66, he and some colleagues had set up the Intermediate Technology Development Group to work on "intermediate" or "appropriate" technologies in a practical way. As the group evolved during the 1970s, a growing emphasis in its work, largely absent from the founder's own writings, was a concern with technologies used by women. In Africa, for example, it was being realized that most food eaten in people's homes was produced by women farmers and gardeners, yet most efforts to improve agricultural technology had been directed at men. In the West, as others pointed out, the design of houses and of domestic equipment was often binding "housewives" more strongly to traditional roles. In both contexts, it could seem that "technical idealism" (as it is described here) was somehow a reflection of male values. Quite separately, too, new socially appropriate approaches to housing and architecture were being expressed.[42]

Yet another strand in thinking about technology during these years arose from a growing awareness that affluent western nations were consuming limited energy resources, especially petroleum, at a rate that could seriously deplete reserves within three or four decades. Even before the oil supply crisis of 1973, a number of informal groups and individuals in Europe, like American enthusiasts for the *Whole Earth Catalog,* were exploring prospects for living in a "self-sufficient" manner, using wind or solar energy to provide their own electricity supplies. Wind turbines (figure 42) became a symbol for what was referred to, from about 1972, as "alternative technology."[43] Soon, however, wind power attracted wider interest and more organized research because the prospect of future energy shortages seemed nearer. Indeed, alternative technology began to seem relevant to official "science policy" in the United States.[44]

Many of these trends reached full tide with the inauguration of Jimmy Carter as president in January 1977. He received Schumacher at the White House, commissioned a major report on the global environment,[45] and initiated policies that encouraged advances in the performance and use of large wind turbines.

Four years later, however, at the start of Ronald Reagan's presidency, oil prices were falling and the energy projects started by the Carter administration were rapidly cut back.[46] Solar energy, wind power, and intermediate technology for Third World countries had taken their place as serious if very minor options, but the sense that they belonged to a distinctive, more socially idealistic type of technology was beginning to dissipate. In Europe, "green" politics now seemed more relevant than alternative lifestyles.

Meanwhile, Reagan gave his support to the revival of a very different kind of idealism in technology. While wind turbines had been attracting so much attention during the 1970s, there had been a continued, very rapid development in electronics and computing, with equipment becoming smaller, cheaper, and more versatile than ever before. One technological possibility that had emerged was the idea of using small but powerful computers to control new weapons that would provide a defense against ballistic missiles. For a long time it had seemed that there could be little warning and no protection against an attack on the United States with such weapons. The only means of prevention was the possession of similar, nuclear-armed missiles to deter the potential aggressor.

However, physicists investigating the structure of matter in a typical "pure science" way had come upon ways of manipulating electrons in their orbits about the atomic nucleus such that very intense and coherent bursts of radiation could be produced. The outcome, in the early 1960s, was first the maser, operating with microwaves such as those used in radar, and then the laser, working with light. By the end of the 1970s many applications of the laser were in regular use, some for printing processes, others in surgery. Powerful laser rays offered the prospect of intercepting and destroying an enemy's ballistic missiles in space, and there were other ideas for using lasers in military equipment.

President Reagan's so-called star wars speech in 1983 announced an ambitious program of research and development involving this and other "high" technologies in a "Strategic Defense Initiative" (SDI). As he presented his scheme, Reagan clearly felt a degree of idealism about having a sure defense against what still seemed a formidable Soviet threat. He even seems to have felt that SDI could rid the world of the possibility of nuclear war.

For the scientists who had proposed the project, however, a different and narrowly technical kind of idealism was also at work. It had been apparent for some years that computers held a special fascination for some individuals who worked with them. Explanations of this often came back indirectly to the point made in earlier chapters about the long history of interest in automatic devices that had some of the characteristics of animate life. The modern computer is a highly developed machine of this kind, which allows its human user to create his or her own "world without people"—a perfectly ordered and rational conceptual world within the computer.[47] But the proposed development of laser weapons, some mounted on satellites, made it possible to extend the notion of creating artificial, inanimate worlds. Long-standing technical interests in rational systems and in automata were thus brought together in a grotesque fulfillment, in a fantasy of battles fought in space between missiles and lasers.

The Hope of Progress

Historians of technology are increasingly providing evidence of invention as an imaginative activity prompted by aesthetic feeling,[48] by symbols and ideals, by interest in abstract rationality, and by what have been termed "archetypal impulses."[49] The question is no longer whether such attitudes have a role, but how they relate to the economically oriented functions of industrial civilization and how we should assess and control them. Some commentators are pessimistic and write about an "unquenchable idealism" underlying an uncontrollable "technological imperative." Others have spoken of certain trends in technology as being partly or wholly a pursuit of vision or even fantasy.[50]

In this book, the term "idealism" has been preferred to "fantasy" because it conveys a greater sense of constructive possibility. It enables one to look for relationships between goals for technological development and social and humanitarian ideals. The danger of not recognizing such positive potential is that advances in technology may seem to lose all meaning; they may appear to be oriented to trivial goals or mindless enthusiasms. Technological progress may then look like a road "from knowhow to nowhere," as one historian of technology has put it.[51]

A similar feeling is expressed in a passage from which the title of this book is taken, where T. S. Eliot admonished the present "wretched generation of enlightened men,"

Betrayed in the mazes of your ingenuities,
Sold by the proceeds of your proper inventions.[52]

We have seen examples of such betrayals in this chapter, notably in the work of architects and scientists in the postwar decades. But that need not always be the fate of social idealism in the practice of technology. In the late 1960s, when pessimism was especially rife, a call for a more positive approach came from Sir Peter Medawar, a leading figure in medical research. Addressing the British Association as its president, he affirmed his belief that human happiness depended on a continuous development in many fields, including science and technology. He declared himself "all in favour of a vigorous critical attitude to technological innovation," but he stressed that "there is all the difference in the world between informed and energetic criticism and a drooping despondency that offers no remedy for the abuses it bewails. . . . To deride the hope of progress is the ultimate fatuity, the last word in poverty of spirit and meanness of mind."[53]

Medawar was one whose sense of direction in science and technology was much influenced by a knowledge of its history, and like others before him he found some of the most helpful guideposts in the writings of Francis Bacon. For there was in Bacon a strong sense of the unity of things; his was a "whole systems" view (in 1960s terminology). There was a respect for nature tempering all he said about how the earth's resources should be exploited. And there was a clear emphasis on the use of knowledge and skill in works of charity and for the public benefit.

"In this present period, we may find Francis Bacon speaking to us more than Descartes the metaphysician-geometer or Galileo the engineer-cosmologist. As deeply as any of his pietistic . . . forerunners, he felt the love of God's creation, the pity for the sufferings of man, and the striving for innocence, humility and charity."[54] He felt that knowledge should be "perfected and governed" in love, and that the fruits of knowledge should be used, not for "profit, or fame or power . . . but for the benefit and use of life."[55]

But it is not sufficient simply to assert a social ideal for technology, whether in Bacon's language, or Bernal's or Schumacher's. It is not enough, either, to demonstrated that existing techniques, properly used, could feed the world's hungry millions or halt pollution of the biosphere. Many postwar tendencies have been toward the development of institutions that are not responsive to such ideals, nor to this concept of "properly using" technology. The need, then, is to open up these institutions to democratic control, backed by relevant public-interest research. Technologists, scientists, and indeed all of us would then have to confront more directly what it might mean to serve society rather than the unpeopled worlds of technical rationality, and hence what it could mean to apply technical knowledge "for the benefit and use of life."

Notes

1. The Cathedral Builders: European Technical Achievement between 1100 and 1280

Useful background material for this chapter will be found in books by Crombie, Fitchen, Gimpel (both works cited in the Select Bibliography), Pevsner, Simson, Norman Smith, and also White.

1. Norman Smith, *A History of Dams,* pp. 91–101 (on Islamic dams) and p. 165 (on Domesday water mills).

2. A. Pacey, *Technology in World Civilization,* 1990.

3. Eric McKee, *Clenched Lap or Clinker,* Greenwich (London): National Maritime Museum, 1972, Introduction. For a more general view of Scandinavian boats and Atlantic navigation, see Magnus Magnusson, *Viking Expansion Westwards,* London: Bodley Head, 1973.

4. Stephen A. Marglin, "What do bosses do?", in *The Division of Labour* ed. André Gorz, Hassocks (England): Harvester Press, 1977, pp. 41–44.

5. This contemporary account of the rebuilding is widely quoted, e.g., by N. Pevsner, *Outline of European Architecture,* chapter 3, pp. 78–79.

6. *The rule of St. Benedict,* ed. and trans. Justin McCann, London: Burns Oates, 1952, caput 48; see also B. J. M. Vignes, in *Saint Bernard et son temps, Congrès de 1927,* Dijon: Association Bourguignonne, 2 vols., 1928, vol. I, pp. 295–317.

7. *The rule of St. Benedict,* caput 57.

8. Measurements from author's fieldwork, but compare David Luckhurst, *Monastic Watermills,* London: Society for the Protection of Ancient Buildings, no date, c. 1963, p. 9. For an example of Cistercian engineering involving a transport canal, see John Weatherill, "Rievaulx Abbey," *Yorkshire Archaeological Journal, 38* (1952–55), pp. 333, 342.

9. J. Gimpel, *The Cathedral Builders,* p. 10, referring to the Aube district or "département."

10. E. M. Carus Wilson, "An industrial revolution of the 13th century," *Economic History Review, 11,* 1st series (1941), pp. 39–60.

11. J. Gimpel, *The Cathedral Builders,* p. 45.

12. N. Pevsner, *Outline of European Architecture,* chapter 3, (pp. 91–92).

13. Watkin Williams, *St. Bernard of Clairvaux,* Manchester: Manchester University Press, 1935, p. 54.

14. R. A. Brown, H. M. Colvin, and A. J. Taylor, *The History of the King's Works,* 2 vol., London: Her Majesty's Stationary Office, 1963, vol. I, p. 156 (on costs) and pp. 204–206 (on Richard L'enginour).

15. Quoted by G. G. Coulton, *Art and the Reformation,* Cambridge (England): Cambridge University Press, 1958, appendix 26.

16. G. G. Coulton, "The high ancestry of Puritanism," in *Ten Medieval Studies,* Cambridge (England): Cambridge University Press, 1930.

17. W. Williams, *St. Bernard of Clairvaux,* p. 142.

2. A Century of Invention: 1250–1350

On medieval clockwork, see Cipolla, Hill, Price, and also Needham and his colleagues, all cited in the Select Bibliography. On maps, drawing, and builders' mathematics, see Bagrow, Harvey, Morgan, and also Simson.

1. J. B. Bury, *The Idea of Progress,* London: Macmillan, 1932, reprinted New York: Dover Books, 1955, p. 27; see also Roger Bacon, *Opus majus,* trans. Robert B. Burke, 2 vol., Philadelphia: Pennsylvania University Press, 1928, pp. 289, 633.

2. Bury, *The Idea of Progress,* p. 28.

3. Emile Mâle, *The Gothic Image,* trans. Dora Nussey, London: Collins, new ed., 1961, p. 20.

4. A. Pacey, *Technology in World Civilization,* pp. 45–51.

5. Some of the evidence for this is given by A. Y. al-Hassan and D. R. Hill, *Islamic Technology,* Cambridge: Cambridge University Press, and Paris: UNESCO, 1986, pp. 184–189.

6.. R. Patterson, "Spinning and weaving," in C. Singer, E. J. Holmyard, A. R. Hall, and T. I. Williams (eds.), *A History of Technology,* vol. 2 (1956), p. 205.

7. Lewis Mumford, *Technics and Civilization,* pp. 12–18.

8. D. J. de S. Price, *On the Origin of Clockwork,* p. 100.

9. Pierre de Maricourt (or Petrus Peregrinus), "Letter on the magnet," part I, trans. Brother Arnold, in *Electrical World, 43* 1904, pp. 598–600.

10. Quoted by Lynn Thorndyke (ed.), *"The Sphere" of Sacrobosco,* Chicago: Chicago University Press, 1949, p. 230.

11. A. C. Crombie, *Robert Grosseteste and the Origins of Experimental Science,* Oxford: Clarendon Press, 1953, pp. 204–208, 210.

12. Vitruvius, *De architectura,* trans. Frank Granger, London: Heinemann, 1931, book I, 3.

13. J. B. A. Lassus (ed.), *Album de Villard de Honnecourt,* Paris, 1858, p. 11; the original reads: "c'est li masons d'on orologe."

14. E. Panofsky, *Abbot Suger,* p. 105.

15. One passage read at the consecration of a new church was Revelation, xxi, 2–5. See Michel Andrieu, *Le pontifical romain au moyen-âge,* vol. I, Vatican City, 1938, pp. 176–195. The twelve foundation stones are referred to in Revelation, xxi, 14.

16. Plato, *Timaeus,* trans. and ed. H. D. P. Lee, Harmondsworth: Penguin, 1965, p. 56.

17. A. C. Crombie, *Robert Grosseteste* (see note 11), pp. 109, 112.

18. Pierre de Maricourt, "Letter on the Magnet," part I (trans. Brother Arnold, pp. 598–600).

19. A. C. Crombie, *Robert Grosseteste* (see note 11), p. 207.

20. Pierre de Maricourt, "Letter on the Magnet," part II (trans. Brother Arnold, pp. 600–601).

21. Quoted by Otto von Simson, *The Gothic Cathedral,* p. 33.

22. J. Gimpel, *The Cathedral Builders,* p. 122; P. Frankl, *The Gothic,* pp. 35–39, 133.

23. Milan cathedral is a later example for which there are records of a controversy about *ad quadratum* design, whose use was rejected; see James S. Ackerman, "ARS SINE SCIENTIA NIHIL EST: Gothic theory of architecture at the Cathedral of Milan," *Art Bulletin, 31,* 1949, pp. 84–111.

24. L. S. Colchester and J. H. Harvey, "Wells Cathedral," *Archaeological Journal, 131,* 1974, p. 214 and fig. 2.

25. B. G. Morgan, *Canonic Design,* pp. 8–67.

26. P. Frankl, *The Gothic,* p. 135.

27. E. Panofsky, *Gothic Architecture and Scholasticism,* New York: Meridian Books, 1951.

28. J. Huizinga, *The Waning of the Middle Ages,* trans. F. Hopman, Harmondsworth: Pelican, 1955, p. 22.

3. Mathematics and the Arts: 1450–1600

For the Renaissance background in technology, see Cipolla, Gille, Keller, and also Parsons in the Select Bibliography.

1. F. D. Praeger, "Brunelleschi's inventions," *Osiris, 9,* 1950, pp. 457–554.

2. T. S. Ashton, *The Industrial Revolution,* new ed., Oxford: Oxford University Press, 1968, p. 11.

3. W. B. Parsons, *Engineers and Engineering in the Renaissance,* p. 106.

4. This is discussed by George Basalla, *The Evolution of Technology,* pp. 129–133.

5. On "movements" in technology, see also A. Pacey, *The Culture of Technology,* pp. 28–29, 86.

6. On the history of guns, see W. Y. Carman, *A History of Firearms,* London: Routledge, 1955.

7. F. Klemm, *A History of Western Technology,* p. 123, and B. Gille, *The Renaissance Engineers,* p. 80.

8. Leone Battista Alberti, *De re aedificatoria,* Florence, 1485; trans. James Leoni as *Ten Books of Architecture,* London, 1726; facsimile ed. London: Tiranti, 1955, book X, chapters 2–12.

9. Alberti, *De re aedificatoria,* book IX, chapter 5.

10. Vincent Cronin, *The Florentine Renaissance,* London: Collins, 1967, pp. 140–142.

11. Norman Smith, *A History of Dams,* p. 128.

12. A. G. Keller, "Pneumatics, automata, and the vacuum in the work of Aleotti," *British Journal for the History of Science, 3,* 1966–67, pp. 338–347.

13. Palladio's book was translated into English by Isaac Ware as *The Four Books of Architecture,* London, 1738; reprinted New York: Dover Books, 1965. For social, economic, and intellectual background, see James S. Ackerman, *Palladio,* Harmondsworth: Pelican, 1966.

14. Vincentio Galilei, *Dialogo . . . della musica antica et della moderna,* Florence, 1581.

15. Stillman Drake, "Renaissance music and experimental science," *Journal of the History of Ideas, 31,* 1970, especially pp. 496–497.

16. Galileo Galilei, *Discorsi e Dimostrazioni Matematiche intorno a Due Nuove Scienze,* Leiden, 1638, "First day", trans. Henry Crew and Alfonso de Salvio as *Dialogues concerning Two New Sciences,* 1914; new ed., New York: Dover Books, no date, c.1960, p. 107.

17. Luigi Bulferetti, *Galileo Galilei nella societa de suo tempo,* Maduria: Lacaito Editore, 1964, pp. 19–22. (I am indebted to the late W. V. Farrar for translating this book.)

18. For a modern edition of Ramelli's book, see *The Various and Ingenious Machines of Agostino Ramelli (1588),* trans. M. T. Gnudi, ed. Eugene S. Ferguson, Baltimore: Johns Hopkins University Press, 1976.

19. A. G. Keller, *A Theatre of Machines,* p. 8.

20. For an example of a book of mechanical fantasies and inventions compiled in a different cultural setting, consider the work written and illustrated in Iraq about the year 1206 by the Islamic technician al-Jazari. See Donald R. Hill, *A History of Engineering in Classical and Medieval Times,* pp. 204–208.

21. These points occur in a book on machines written by Galileo about 1600 and published by Stillman Drake and I. E. Drabkin in their volume *Galileo Galilei: On Motion and On Mechanics,* Madison: Wisconsin University Press, 1960.

22. Drake and Drabkin (eds.), *Galileo Galilei: On Motion and On Mechanics,* p. 150.

23. Raymond de Roover, "A Florentine firm of cloth manufacturers," *Speculum, 16,* 1941, 1–33.

24. Quoted by F. Klemm, *A History of Western Technology,* p. 163.

25. Galileo Galilei, *Discorsi e Dimostrazioni,* "Second day," trans. Henry Crew (as above), pp. 115–117.

26. Galileo Galilei, *Discorsi e Dimostrazioni,* "Second day," p. 144.

27. Anthony Blunt, *Philibert de l'Orme,'* London: Zwemmer, 1958, p. 114.

28. Mark Girouard (ed.), "The Smythson Collection of the Royal Institute of British Architects," *Architectural History, 5,* 1962, pp. 72, 79, 112, 119, etc.

29. Evidence for this statement is offered in chapter 7 (where notes 10–15 cite the relevant sources). I am indebted to Jerry Ravetz for some ideas about Galileo; see his "Galileo on the strength of materials," *School Science Review, 149,* 1961, pp. 52–57.

4. The Practical Arts and the Scientific Revolution

Background material for this chapter can be found in Hall, Kearney, Merchant, Wolf; also Pacey, *Technology in World Civilization,* chapter 6.

1. For more details of these inventions, see A. G. Keller, "Pneumatics, automata and the vacuum in the work of Aleotti," *British Journal for the History of Science, 3,* 1966–67, pp. 338–347; also C. F. Dendy Marshall, "Some mechanical inventions," *Newcomen Society Transactions, 16,* 1937, pp. 1–26.

2. Milton Kerker, "Science and the steam engine," *Technology and Culture, 2,* 1961, p. 383.

3. Otto von Guericke, *Experimenta nova Magdeburgica de vacuo spatio,* 1672; facsimile ed., introduction by A. Rosenfeld and O. Zeller, Aalen, 1962.

4. Luigi Bulferetti, *Galileo Galilei nella societa de suo tempo,* Manduria: Lacaito Editore, 1964, p. 25.

5. Edmé Mariotte, *Traité du mouvement des eaux et des autres corps fluides,* Paris, 1686, trans. J. T. Desaguliers as *A treatise of Hydrostaticks,* London, 1718.

6. Antoine Parent, "Sur la plus grande perfection possible des machines dont un fluide est la force mouvente," *Histoire et Mémoires de l'Académie Royale des Sciences,* 1704; 2nd ed., Paris, 1722, *Mémoires,* pp. 323–336.

7. On Parent and Papin, see G. R. Talbot and A. J. Pacey, "Antecedents of thermodynamics in the work of Guillaume Amontons," *Centaurus, 16,* 1971, pp. 20, 23–27.

8. This letter and other documents on Papin and Huygens are printed by F. Klemm, *A History of Western Technology,* pp. 213–227.

9. Evidence for this is discussed in the much underrated book by L. T. C. Rolt, *Thomas Newcomen: the Prehistory of the Steam Engine,* Newton Abbot: David & Charles, 1963.

10. Quoted by A. E. Musson in his introduction to the 1963 ed. of H. W. Dickinson, *A Short History of the Steam Engine.*

11. Rolt, *Thomas Newcomen,* makes this point, but I am also indebted to W. N. Slatcher for stressing its significance.

12. Allan Brocket, *Nonconformity in Exeter, 1650–1875,* Manchester: Manchester University Press, 1962, pp. 59–60, 68.

13. A. Rupert Hall, "Engineering and the scientific revolution," *Technology and Culture, 2,* 1961, p. 336. The same point is made with great clarity by George Basalla, *The Evolution of Technology,* pp. 95–97.

14. Engineers and others who used Galileo's theory around 1800 include William Reynolds, Charles Bage, John Banks, and probably Thomas Telford; see chapter 7.

15. *Architecture hydraulique* was written by Forest de Belidor and published in four volumes, Paris, 1737–53.

16. P. J. Booker, "Gaspard Monge and his effect on engineering drawing," *Newcomen Society Transactions, 34,* (1961–62), pp. 15–36.

17. For a useful discussion of differing views on the influence of science, see A. E. Musson and Eric Robinson, *Science and Technology in the Industrial Revolution,* pp. 26–30.

18. René Descartes, *Principia philosophiae,* 1644, iv, 203; in *Oeuvres de Descartes,* ed. Adam and Paul Tannery, vol. 8, Paris, 1905, p. 326.

19. René Descartes, *Traité de l'homme,* in *Oeuvres de Descartes,* ed. Adam and Paul Tannery, vol. 11, Paris, 1909, pp. 120, 130–131.

20. Quoted by Charles Webster, "Henry Power's experimental philosophy," *Ambix* (Journal of the Society for the study of Alchemy and early Chemistry), *14,* 1967, p. 176.

21. Lucretius, *The Nature of the Universe,* trans. Ronald Latham, Harmondsworth: Penguin, 1951, p. 18.

22. *Opere di Galileo Galilei,* ed. Antonio Favaro, Edizione Nationale, Florence, 1890–1909, vol. 6, pp. 347–348.

23. Antoine Parent, *Essais et recherches de mathématique et de physique,* 2nd ed., 3 vol., Paris, 1713, vol. 1, part 2.

24. Galileo Galilei, *Discorsi e Dimostrazioni Matematiche intorno a Due Nuove Scienze,* Leiden, 1638, "First day", trans. Henry Crew and Alfonso de Salvio as *Dialogues concerning Two New Sciences,* 1914; new ed., New York: Dover Books, no date, c.1960, p. 94.

25. John Stuart Mill, "Essay on Bentham," in F. R. Leavis (ed.), *Mill on Bentham and Coleridge,* London: Chatto & Windus, 1950.

26. Quoted by E. Strauss, *Sir William Petty, Portrait of a Genius,* London: Bodley Head, 1954, p. 52.

27. This and quotations in the previous two paragraphs are from Strauss, *Sir William Petty,* pp. 61–63.

28. A. Pacey, *Technology in World Civilization,* pp. 97–102.

5. Social Ideals and Technical Change: German Miners and English Puritans, 1450–1650

On metals and mining, see Adams, Agricola (Hoover ed.), Aitchison, also Donald in the Select Bibliography. On the economic, social, and religious background, see Rich and Wilson (ed.), then Tawney, also Webster's *The Great Instauration.*

1. Figures from the bicentenary publication of the Freiberg Mining Academy, *Bergakademie Freiberg Festschrift,* 1765–1965, Freiberg, 1965, p. 9.

2. Quoted by C. M. Cipolla, *European Culture and Overseas Expansion,* pp. 39–40.

3. Earl J. Hamilton, *American Treasure and the Price Revolution in Spain, 1501–1650*. Cambridge, MA: Harvard Economic Studies, 1934, p. 35; on inflation in Saxony, see pp. 206–209.

4. Another major source of metal was Sweden. Output of copper (and some silver) increased from the 1570s onward as the German mines faced economic difficulties, the most important mine complex being at Falun, northwest of Stockholm. On European inflation and the devaluation of some currencies by reducing the silver content of coins, see F. P. Braudel, "Prices in Europe," in *Cambridge Economic History of Europe*, ed. E. E. Rich and C. H. Wilson, vol. 4, Cambridge: Cambridge University Press, 1967, pp. 374–486.

5. M. B. Donald, *Elizabethan Copper,* p. 202. The units of measurement used here are 1 cs. = 50 kg; 1 lb = 453 g; 1 lot = 14 g.

6. Donald, *Elizabethan Copper,* p. 206. Note that 4/11 means 4 shillings and 11 pence, equal to £0.24, and 30/- is £1.50. There were 240d. (pence) or 20/- in £1.00.

7. For documents of 1565 and 1602 describing the wheel, see Donald, *Elizabethan Copper,* pp. 166, 169. For the recent survey, see Christopher D. Jones, "Goldscope Mine—a short history," *The Mine Explorer* (Cumbrian Amenity Trust), vol. 1, 1984, pp. 29–34.

8. The trough was mentioned in 1565 but not the dam. However, there is still an old earth dam on the site, although probably later than 1565. It measures 22 or 23 feet (6.8 meters) high and has pitched-stone facing (author's fieldwork). Levels within the mine indicating how water may have flowed from the waterwheel are illustrated by John Postlethwaite, *Mines and Mining in the English Lake District,* London, 1877, new ed., Beckermeet: Michael Moon, 1975, p. 77.

9. Georgius Agricola, *De re metallica,* trans. Herbert Clark Hoover and Lou Henry Hoover, p. 23.

10. L. Aitchison, *A History of Metals,* vol. 1, p. 292.

11. Place associated with Elijah in I Kings xvii, 9–10 and Luke, iv, 26.

12. Agricola, *De re metallica,* pp. 99, 118.

13. Quoted by E. G. Schwiebert, *Luther and his Times,* St. Louis (Missouri): Concordia Publishing, 1950, pp. 663–667, 675.

14. Quoted by F. D. Adams, *The Birth and Development of the Geological Sciences,* p. 199; I am indebted to Sylvia Farrar for translating the verse.

15. E. G. Schwiebert, *Luther and his Times,* pp. 405–406.

16. Quoted by Adams, *The Birth and Development of the Geological Sciences,* p. 198.

17. A. E. Waite (ed.), *The Hermetical and Alchemical Works of Paracelsus,* 2 vols., London, 1894, vol. 1, pp. 72–73.

18. *The Hermetical and Alchemical Works of Paracelsus,* vol. 1, p. 20.

19. Agricola, *De re metallica,* trans. Hoover (see note 9), pp. x–xi.

20. John U. Nef, *Industry and Government in France and England, 1540–1640,* Philadelphia: American Philosophical Society, 1940, pp. 1, 98.

21. Charles Webster, *The Great Instauration,* p. 354.

22. R. H. Tawney, *Religion and the Rise of Capitalism,* Harmondsworth: Pelican, 1938, reprinted, 1964. See the acknowledgment to Max Weber, *The Protestant Ethic and the Spirit of Capitalism,* 1930, on pp. ix–x, 311–313n. On Calvinism in Geneva and Massachusetts, and on the later development of an individualistic ethic, see Tawney pp. 124–125, 135–138, 232–233.

23. F. Klemm, *A History of Western Technology,* pp. 191–197.

24. Hugh Trevor-Roper, *Religion, the Reformation and Social Change,* London: Macmillan, 1967, p. 46.

25. K. F. Helleiner, "The population of Europe," in *Cambridge Economic History of Europe,* vol. 4, ed. E. E. Rich and C. H. Wilson, Cambridge: Cambridge University Press, 1967, p. 41.

26. F. Klemm, *A History of Western Technology,* pp. 186–188; Charles Webster, *The Great Instauration,* pp. 386–387.

27. On Bacon's *Instauratio Magna* and ideas about the millennium, see Charles Webster, *The Great Instauration,* pp. 21, 23, 27; also see Webster's *From Paracelsus to Newton,* pp. 30–31, 51.

28. The "new logic" was published as *Novum Organum* in 1620. Only two parts of the natural history were published, in 1622 and 1623.

29. Francis Bacon, *The Advancement of Learning* book I, *1,* 3, in *The Advancement of Learning and New Atlantis,* ed. Arthur Johnston, Oxford: Clarendon Press, 1974.

30. Francis Bacon, *Sylva Sylvarum,* 1626; see the comments of Charles Webster, *From Paracelsus to Newton,* pp. 61–62.

31. John Webster, *Academiarum examen,* London, 1654, pp. 5, 12, 14.

32. Webster, *Academiarum examen,* p. 19.

33. Webster, *Academiarum examen,* pp. 105–106.

34. For the mines visited by Webster, see M. C. Gill, *The Yorkshire and Lancashire Lead Mines,* Sheffield: Northern Mines Research Society, n.d., c.1986, pp. 6, 39.

35. Cyril Stanley Smith, *A History of Metallography,* 1960, p. xviii; also C. S. Smith, "Art, technology and science: notes on their historical interaction," *Technology and Culture, 11,* 1970, pp. 493–549.

36. C. H. Wilson, "Trade, society, and the state," in *Cambridge Economic History of Europe,* vol. 4, ed. E. E. Rich and C. H. Wilson, Cambridge: Cambridge University Press, 1967, p. 490.

37. R. H. Tawney, *Religion and the Rise of Capitalism,* pp. 238, 313.

6. The State and Technical Progress: 1660–1770

Background reading indicated in the Select Bibliography includes Artz, Chambers and Mingay, Cobban, Cole, Jensen, and also C. T. Smith. Ideas and information have also come from discussions with David M. Farrar and his unpublished thesis, *The Royal Hungarian Mining Academy,* M.Sc. thesis, University of Manchester Institute of Science and Technology, 1971.

1. Quoted by F. Klemm, *A History of Western Technology,* p. 219.

2. A. E. Bell, *Christian Huygens and the Development of Science in the Seventeenth Century,* London: Edward Arnold, 1947, p. 61.

3. C. W. Cole, *Colbert,* vol. 1, pp. 360–362.

4. Cole, *Colbert,* vol. 1, p. 381.

5. J. S. Allen, "The introduction of the Newcomen engine, 1710–1733," *Newcomen Society Transactions, 42,* 1969–70, pp. 169–190.

6. F. Williamson, "George Sorocold of Derby: a pioneer of water supply," *Journal of the Derbyshire Archaeological and Natural History Society,* 1936, p. 43.

7. Daniel Defoe, *A Tour through the Whole Island of Great Britain (1724–6),* London: Dent, 2 vols., 1962; Celia Fiennes, *Journeys,* ed. Christopher Morris, London: Crescent Press, 1947.

8. "Epistle to Burlington," lines 197–204, *The Poems of Alexander Pope,* ed. J. E. Butt, London: Methuen, and New Haven, CT: Yale University Press, Twickenham ed., 1939–61, vol. 3(ii), p. 151.

9. Howard Erskine Hill, *The Social Milieu of Alexander Pope,* New Haven, CT: Yale University Press, 1975, p. 30.

10. Erskine Hill, *The Social Milieu of Alexander Pope,* pp. 219, 230–231.

11. J. T. Desaguliers, *A Course of Experimental Philosophy,* 2 vols, 1734, 1744, 3rd ed., London, 1763, vol. 1, p. 283. Ralph Allen's wagonway is illustrated by M. J. T. Lewis, *Early Wooden Railways,* London: Routledge, 1970, plates 56, 62, 63.

12. Mary Chandler, *Poems on Several Occasions,* London, 1734; also quoted by F. D. Klingender, *Art and the Industrial Revolution,* p. 68.

13. Pope, "Epistle to Burlington," lines 185–188, *The Poems of Alexander Pope,* (as cited in note 8), vol. 3(ii), pp. 150–151.

14. C. W. Cole, *Colbert,* vol. 1, pp. 314, 381.

15. Pope, "Epistle to Addison," lines 53–54, *The Poems of Alexander Pope* (as cited in note 8), vol. 6, p. 204.

16. C. W. Cole, *Colbert,* vol. 1, p. 24.

17. John Irwin and Katharine Brett, *Origins of Chintz,* London: Her Majesty's Stationary Office, 1970, p. 1.

18. A. Pacey, *Technology in World Civilization,* chapters 7 and 8.

19. John U. Nef, *Industry and Government in France and England, 1540–1640,* Philadelphia: American Philosophical Society, 1940, pp. 2, 98.

20. Regulation of the French dyeing industry was introduced by Colbert in 1667 and was reformed in 1731. Directors or inspectors of dyeworks between the latter date and the revolution were, successively, Dufay, Hellot, Macquer, and Berthollet. See K. G. Ponting, *A Dictionary of Dyes and Dyeing,* London: Mills, 1980; and A. Juvet-Michel, "Textile printing in 18th-century France," *CIBA Review* (Basel), No. 31, March 1940, pp. 1091–1099. I am indebted to Shirley Armer for suggestions about this and previous paragraphs.

21. C. T. Delius, *Anleitung zu der Bergbaukunst,* Vienna, 1773; trans. J. G. Schreiber as *Traité sure la science de l'exploitation des mines,* 2 vols., Paris, 1778. See vol. 2, pp. 323 etc., in the French edition. Another key text on mining in

Europe is Héron de Villefosse, *De la richesse minérale,* 3 vols and atlas, Paris, 1819; I have drawn on volume 1 here.

22. C. A. Macartney, *The Habsburg Empire, 1790–1918,* London: Weidenfeld and Nicolson, 1968, pp. 48–49; also Henry Marczali, *Hungary in the Eighteenth Century,* Cambridge: Cambridge University Press, 1910, p. 319.

23. Schemnitz is now within the borders of Czechoslovakia and is known as Banskà Stiavnica.

24. J. F. Weidler, *Tractatus de machinis hydraulicis,* Wittenberg, 1728; 2nd ed., 1735, section iv, p. 91.

25. C. W. Cole, *Colbert,* vol. 1, p. 315.

26. J. B. Fischer von Erlach, *A Plan of Civil and Historical Architecture,* trans. T. Lediard, London, 1737.

27. Pope, "Epistle to Burlington," lines 179–80; *The Poems of Alexander Pope,* vol. 3 (ii), p. 154.

28. Rev. John Dalton, *Descriptive Poem Addressed to Two Ladies at their return from viewing the Mines near Whitehaven,* London, 1755, p. 12.

7. Technology in the Industrial Revolution

For other aspects of the industrial revolution in Britain, see Select Bibliography citations for Ashton, Fitton and Wadsworth, the three books by Hills, also Hyde, then Klingender, and finally, Musson and Robinson. A quite different but complementary discussion is to be found in Pacey, *Technology in World Civilization,* chapters 6 and 7.

1. Arthur Raistrick, "The steam engine on Tyneside," *Newcomen Society Transactions, 17* (1936–37), pp. 153–154.

2. Earlier footbridges made of iron were near Middleton-in-Teesdale, County Durham (a suspension bridge of 1741), and at Kirklees Hall, Yorkshire (1770; information from David Nortcliffe).

3. Neil Cossons and Barrie Trinder, *The Iron Bridge: symbol of the Industrial Revolution,* Bradford-on-Avon: Moonraker Press, 1979, p. 11.

4. Evidence of J. Strutt, Select Committee Report on Children in Manufactories, *Parliamentary Papers,* 1816, *3,* p. 235.

5. Quoted by L. T. C. Rolt, *James Watt,* London: Batsford, 1962, p. 86.

6. On India, see Pacey, *Technology in World Civilization,* pp. 117–121.

7. R. L. Hills and A. J. Pacey, "The measurement of power," *Technology and Culture, 13* (1972), pp. 29–31.

8. A. E. Musson and Eric Robinson, *Science and Technology in the Industrial Revolution,* pp. 407–411, 413–423.

9. A. W. Skempton and H. R. Johnson, "The first iron frames," *Architectural Review, 131* (1962), pp. 175–186.

10. An account of Reynolds's tests was given by John Banks, *On the Power of Machines,* Kendal, 1803, pp. 88–90; data from the tests are quoted by Thomas Tredgold, *Practical Essays on the Strength of Cast Iron,* 2nd ed., London, 1824, p. 79.

11. Discussed by A. W. Skempton, "The origin of iron beams," *Actes du VIIIe Congres International d'Histoire des Sciences,* Florence, 1956, p. 1037.

12. Charles Bage, "Theorems of the Strength of Iron," ms. appended to a letter from Bage to William Strutt, 29 August 1803, Bage papers, Shrewsbury Public Library.

13. William Emerson, *The Principles of Mechanics,* London, 1758, 4th ed., 1794, p. 93, etc.

14. Olinthus Gregory, *A Treatise of Mechanics,* 2 vols. and plates, London, 1806, vol. I, p. 104.

15. Banks, *On the Power of Machines,* pp. 91–103.

16. "Lee's experiments with indicator," ms. in box 26, Boulton and Watt Collection, Birmingham Reference Library, discussed by A. J. Pacey, "Some early heat engine concepts," *British Journal for the History of Science, 7* (1974), pp. 135–143.

17. Hills and Pacey, "The measurement of power," pp. 25–28.

18. For a stimulating review of the debate among historians pointing out that "scientific attitudes were much more widespread . . . than scientific knowledge," see Peter Mathias, "Who unbound Prometheus?," in P. Mathias (ed.), *Science and Society, 1600–1900,* Cambridge: Cambridge University Press, 1972, pp. 54–80.

19. Barrie Trinder, *The Industrial Revolution in Shropshire,* pp. 199–200.

20. John Smeaton, *Experimental Enquiry,* London, 1796. This consists of papers previously published in the *Philosophical Transactions of the Royal Society,* 1759, 1776, and 1782, of which the paper on waterwheels (and windmills) is the first, and had previously been reprinted in 1760. It was reprinted again in Smeaton's *Miscellaneous Papers,* London, 1814.

21. Smeaton, *Experimental Enquiry,* 1796, p. 17.

22. Isaac Newton, *Principia,* 1687, book III, 4th rule of reasoning; see Florian Cajori (ed.), *Sir Isaac Newton's Mathematical Principles* of Natural Philosophy, Andrew Motte's trans., 1729, as edited 1934. University of California Press, 1962.

23. *Reports of the late John Smeaton, FRS,* 3 vols., London, 1812, preface, vol. 1, pp. iii–xi.

24. S. B. Donkin, "The Society of Civil Engineers (Smeatonians)," *Newcomen Society Transactions, 17* (1936–37), pp. 51–64. See also Denis Smith, "Professional practice," in A. W. Skempton (ed.), *John Smeaton, FRS,* London: Thomas Telford, 1981.

25. Eugene S. Ferguson, *Kinematics of Mechanism from the time of Watt,* United States National Museum, Bulletin 228, Washington, D.C., 1962.

26. *Journal de l'Ecole Polytechnique, 1,* 3rd cahier, year 4 (1796), new regulations for the "stereometry year" in the course.

27. P. J. Booker, "Gaspard Monge and his effect on engineering drawing," *Newcomen Society Transactions, 34* (1961–62), pp. 15–36.

28. P. L. Lanz and Augustin de Bétancourt, *Essai sur la composition des machines,* 2nd ed., Paris, 1819, opening sentence.

29. D. S. Landes, *The Unbound Prometheus,* p. 24.

30. Stuart Smith, *A View from the Iron Bridge,* Ironbridge (England): Ironbridge Gorge Museum Trust; also F. D. Klingender, *Art and the Industrial Revolution,* 1947, Paladin ed., 1972, pp. 75–78; and Cossons and Trinder, *The Iron Bridge,* pp. 58–59.

31. Alexander Pope, "Epistle to Burlington," and Rev. John Dalton, "Descriptive poem." For context and full citation, see chapter 6 and notes 8 and 28 to that chapter.

32. Mary Dixon, "Letter to the committee of civil engineers," in *Reports of the late John Smeaton, FRS,* 3 vols., London, 1812, vol. 1, pp. xxv–xxx.

33. Quoted by B. Trinder, *The Industrial Revolution in Shropshire,* 1973, p. 199; see also Rachel Labouchere, *Abiah Darby,* York (England): Sessions, 1988.

34. Arthur Raistrick, *Dynasty of Iron Founders: the Darbys and Coalbrookdale,* Newton Abbot: David and Charles, 1970, pp. 66, 91–92.

35. Adam Smith, *An Inquiry into the Nature and Causes of the Wealth of Nations,* London, 1776; Pelican ed., ed. Andrew Skinner, Harmondsworth: Penguin, 1970, book I, chapter 2.

36. Joseph Priestley, *An Essay on the First Principles of Government,* London: Dodsley, Cadell and Johnson, 1768, pp. 131, 137.

37. Ms. letter, George Lee to James Watt, Jr., September 1793, Boulton and Watt Collection, Birmingham Reference Library.

38. Ms. letters, Charles Bage to William Strutt, 1801–1818, Shrewsbury Public Library, mss. 2657. (These letters discuss iron roofs briefly and bleaching at length, ending on 25 February, 1818, with Strutt optimistic and Bage pessimistic about, "the improving state of Man in Society.")

39. John Banks, *A Treatise on Mills,* London and Kendal, 1795, 2nd ed., 1815, preface.

40. Letters written to James Watt by his wife Annie, daughter of James McGrigor, are quoted by A. E. Musson and Eric Robinson, *Science and Technology in the Industrial Revolution,* pp. 279–281.

41. Mary Mack, *Jeremy Bentham: an Odyssey of Ideas 1748–1782,* London: Heinemann, 1962, pp. 102–103. Priestley, *An Essay on the First Principles of Government,* p. 17.

42. Michael Roberts, *The Whig Party, 1807–12,* London: Macmillan, 1939, p. 264; compare Elie Halévy, *History of the English People in the Nineteenth Century,* trans. E. I. Watkins and D. I. Barker, 2 vols., new ed., London: Benn, 1949, vol. I, p. 586.

43. Mary Mack, *Jeremy Bentham,* pp. 296–298.

44. Stephen A. Marglin, "What do bosses do?", in *The Division of Labour,* ed. André Gorz, Hassocks (England): Harvester Press, 1977, pp. 13–41.

45. C. L. Hacker, "William Strutt of Derby," *Journal of the Derbyshire Archaeological and Natural History Society, 80* (1960), pp. 55–60.

46. Michael C. Egerton, "William Strutt and the application of convection to the heating of buildings," *Annals of Science, 24* (1968), pp. 73–87.

47. Ms. letter, Jeremy Bentham to William Strutt, 17 July 1794, Strutt papers, Derby Public Library.

48. Ms. letter, Samuel Bentham to William Strutt, 20 July 1805, Strutt papers, Derby Public Library.

8. Conflicting Ideals in Engineering: America and Britain, 1790–1870

Key books for this chapter are those by Hindle, Hounshell, Hunter, Rosenberg, and also Saul; and for Britain, Binnie, Finer, and also Rolt (all three titles cited).

1. H. J. Hopkins, *A Span of Bridges,* Newton Abbot (England): David and Charles, 1970; also Charles E. Peterson, "Iron in early American roofs," *Smithsonian Journal of History, 3,* 1968, section on Thomas Paine and bridges, pp. 47–52.

2. Louis C. Hunter, *Steamboats on the Western Rivers,* p. 33. Hunter's comparison is with the British Empire; the figure of 10,000 tons allows for steamboats on French and German rivers also.

3. Brooke Hindle, *Emulation and Invention,* preface.

4. Cyril Stanley Smith, *A History of Metallography,* Chicago: Chicago University Press, 1960.

5. Hunter, *Steamboats on the Western Rivers,* p. 27.

6. Nathan Rosenberg, "Technological change in the machine-tool industry, 1840–1910," *Journal of Economic History, 23,* 1963, p. 414.

7. Andrew Ure, *The Philosophy of Manufactures,* London, 1835, p. 37.

8. A. E. Musson and Eric Robinson, *Science and Technology in the Industrial Revolution,* pp. 473, 478.

9. Carolyn C. Cooper, "The Portsmouth system of manufactures," *Technology and Culture, 25,* 1984, pp. 182–216.

10. Paul Clements, *Marc Isambard Brunel,* London: Longman, 1970; Samuel Bentham, as Inspector of Naval Works, also had a role in the block-making project. See Carolyn C. Cooper, "The production line at Portsmouth Block Mill," *Industrial Archaeology Review, 6,* 1981–82, pp. 28–44.

11. Merritt Roe Smith, "Army ordnance and the 'American system'," in M. R. Smith (ed.), *Military Enterprise and Technological Change,* Cambridge, MA: MIT Press, 1985, pp. 39–47.

12. Robert S. Woodbury, "The legend of Eli Whitney and interchangeable parts," *Technology and Culture, 1,* 1959–60, pp. 235–253.

13. Many fresh insights in other works cited here stem from Merritt Roe Smith, *Harpers Ferry Armory and the New Technology,* Ithaca: Cornell University Press, 1977.

14. S. B. Saul, "The market and the development of the mechanical engineering industries," in S. B. Saul (ed.), *Technological Change: the U.S. and Britain in the Nineteenth Century,* pp. 146–147.

15. Whitworth's report is in N. Rosenberg (ed.), *The American System of Manufactures,* Edinburgh: Edinburgh University Press, 1969.

16. Howard L. Blackmore, "Colt's London Armoury," in S. B. Saul (ed.), *Technological Change,* pp. 171–195.

17. E. Ames and N. Rosenberg, "The Enfield Arsenal," in S. B. Saul (ed.), *Technological Change,* pp. 100–116.

18. David A. Hounshell, *From the American System to Mass Production,* pp. 4, 19–21, 27.

19. Colt's evidence in: "Reports from committees . . . Small Arms," *Parliamentary Papers,* 1854, *18,* pp. 84–99.

20. Ames and Rosenberg, "The Enfield Arsenal," p. 116.

21. David A. Hounshell, *From the American System to Mass Production,* p. 91.

22. The quotations in this and the previous paragraph are all from Andrew Ure, *The Philosophy of Manufactures,* London, 1835, p. 1.

23. Andrew Ure, *Dictionary of Arts, Manufactures and Mines,* 1st ed., London, 1839; 4th ed., London 1853, article "Spinning," vol. 2, p. 695.

24. Ure, *The Philosophy of Manufactures,* p. 23.

25. Ure, *The Philosophy of Manufactures,* pp. 8, 301.

26. Ms. letter, Ewart to Babbage, 4 March 1824, British Museum (British Library), additional mss., 37183, f. 110.

27. For the relationship of the ideas of Ure, Babbage, and Marx, see R. S. Rosenbloom, "Men and machines," *Technology and Culture, 5,* 1964, pp. 491–496.

28. Quoted from T. B. Bottomore (ed.), *Karl Marx—Early Writings,* London: Watts, 1963, p. 170.

29. Stephen A. Marglin, "What do bosses do?" in André Gorz (ed.), *The Division of Labour,* Hassocks (England): Harvester Press, 1977, pp. 13–41.

30. David F. Noble, "Automation madness," in Steven L. Goldman (ed.), *Science, Technology and Social Progress,* Bethlehem: Lehigh University Press, 1989, p. 75.

31. Merritt Roe Smith, "Army ordnance and the 'American system' " (as cited in note 11), p. 85.

32. C. L. Hacker, "William Strutt of Derby," *Journal of Derbyshire Archaeological and Natural History Society, 80,* 1960, pp. 57–59. This author does not mention Strutt's round "Panoptican" factory, but note that Samuel Bentham had a hand in its design; that and his contribution to the Portsmouth block-making workshops indicate Bentham's strong interest in "rational factory organization." Carolyn C. Cooper, "The production line at Portsmouth Block Mill," *Industrial Archaeology Review, 6,* 1981–82, p. 28.

33. Edwin Chadwick, *The Sanitary Conditions of the Labouring Population of Great Britain,* London, 1842: reprint, ed. M. W. Flinn, Edinburgh: Edinburgh University Press, 1965, pp. 423–424.

34. J. W. Sellek and D. M. Farrar, "Urban sewer renewal," paper read to the Institution of Municipal Engineers at Washington, County Durham, 14 February 1980 (cyclostyled).

35. G. M. Binnie, *Early Victorian Water Engineers,* pp. 12, 33, 130–131.

36. S. E. Finer, *The Life and Times of Sir Edwin Chadwick,* p. 449.

37. J. F. LaTrobe Bateman, speech on completion of Glasgow waterworks, 1859, quoted by Peter E. Russell, *J. F. LaTrobe Bateman,* M.Sc. thesis, University of Manchester Institute of Science and Technology, 1980.

38. Illustrated by L. T. C. Rolt, *Victorian Engineering,* London: Allen Lane, 1970, plate 20.

39. Finer, *The Life and Times of Sir Edwin Chadwick,* pp. 439–440. Although much engineering work was in progress and completed during 1860–90, Chadwick found "progress made . . . bitterly disappointing" (p. 501).

40. R. Fremdling, "Railroads and German economic growth," *Journal of Economic History, 37,* 1977, pp. 583–601; also Dorothy R. Adler, *British Investment in American Railways, 1833–52,* ed. Muriel E. Hidy, Charlottesville: University Press of Virginia, 1970.

41. John F. Stover, *Iron Road to the West: American railroads in the 1850s,* New York: Columbia University Press, 1978, p. 194.

42. David B. Steinman and Sara Ruth Watson, *Bridges and their Builders,* New York: G. P. Putnam's Sons, 1941, p. 207; also David P. Billington and Robert Mark, "The cathedral and the bridge: structure and symbol," *Technology and Culture, 25,* 1984, pp. 37–52.

43. C. W. Condit, "Another view of 'The cathedral and the bridge,' " *Technology and Culture, 25,* 1984, pp. 589–601.

44. Neil Cossons and Barrie Trinder, *The Iron Bridge: symbol of the Industrial Revolution,* Bradford-on-Avon: Moonraker Press, 1979, pp. 53, 58–59, etc.

45. Quoted by Steinman and Watson, *Bridges and their Builders,* pp. 246–267.

46. Hunter, *Steamboats on the Western Rivers,* pp. 102–103, 141.

47. Quoted by Steinman and Watson, *Bridges and their Builders,* p. 220.

48. Quoted by Binnie, *Early Victorian Water Engineers,* p. 36.

49. S. B. Saul, "The market and the development of the mechanical engineering industries," p. 170.

50. Bruce Mazlish (ed.), *The Railroads and the Space Program,* Cambridge, MA, 1965, pp. 8, 53–56, 208; and A. Pacey, *Technology in World Civilization,* pp. 162–164.

51. For the debate among economic historians, see H. J. Habakkuk, *American and British Technology in the Nineteenth Century,* New York: Cambridge University Press, 1962; S. B. Saul (ed.), *Technological Change;* Nathan Rosenberg, *Technology and American Economic Growth.*

9. Institutionalizing Technical Ideals, 1800–1920

For more background on institutions discussed in this chapter, see books listed in the Select Bibliography by Buchanan, Bud and Roberts, Hounshell, Morison, Rossiter, also Smith and Wise. For discussion of early military-industrial complexes, see the book edited by Merritt Roe Smith, also W. H. McNeill.

1. L. T. C. Rolt, *Isambard Kingdom Brunel,* London: Longman, 1957; 7th impression, 1971, pp. 322–323.

2. Compare "technological virtuosity" in company policies as discussed by J. K. Galbraith, *The New Industrial State,* 2nd British edition, London: André Deutsch, 1972, chapter 15.

3. This phrase is derived from Jon Moris's view of modern irrigation schemes as "privileged" technology in his article, "Irrigation as a privileged solution in African development," *Development Policy Review, 5,* (1987), pp. 99–123.

4. George Basalla, in *The Evolution of Technology* (1988), pp. 70, 144–158, portrays economic influences as part of a selection process that decides which inventions are developed.

5. Cecil O. Smith, "The longest run: public engineers and planning in France," *American Historical Review, 95* (1990), pp. 657–692.

6. Smith, "The longest run," p. 675; on American railroad bridges, see, for example, John F. Stover, *Iron Road to the West,* New York: Columbia University Press, 1978, pp. 191–195.

7. Smith, "The longest run," p. 676.

8. Joseph Harriss, *The Eiffel Tower,* London: Paul Elek, 1976, pp. 10–15.

9. *Journal de l'École Polytechnique,* vol. 1, 2nd cahier, year 4 (1796), "Avant-propos."

10. R. A. Buchanan, *The Engineers,* 1989, pp. 161–165.

11. Brooke Hindle, *Emulation and Invention,* 1981, pp. 85–98.

12. Margaret W. Rossiter, *The Emergence of Agricultural Science,* 1975, pp. 25–26.

13. E. John Russell, *A History of Agricultural Science in Great Britain,* London: Allen & Unwin, 1966, pp. 91–94.

14. Tony Travis, "Early intermediates for the synthetic dyestuffs industry," *Chemistry and Industry,* 15 August 1988, pp. 508–509. I am also indebted to Judith Brown and Elizabeth Rigby for information on this subject, and for pointing out material in Maurice R. Fox, *Dyemakers of Great Britain, 1856–1915,* Manchester: Imperial Chemical Industries, 1987.

15. Elting E. Morison, *From Know-how to Nowhere,* 1974, p. 115.

16. Lillian Hoddeson, "The emergence of basic research in the Bell telephone system, 1875–1915," *Technology and Culture, 22,* 1981, pp. 512–544.

17. R. Bud and G. K. Roberts, *Science versus Practice: Chemistry in Victorian Britain,* 1984, pp. 75, 95.

18. The Scottish chemist Alexander Crum Brown also made contributions to ideas about bonds and molecular structures, and the story of Kekulé's work is altogether more involved than indicated here. See Colin A. Russell, *The History of Valency,* Leicester: Leicester University Press, 1971, pp. 98–102.

19. Edwin Layton, "Mirror image twins: the communities of science and technology in 19th-century America," *Technology and Culture, 12* (1971), pp. 562–580.

20. Crosbie Smith and M. Norton Wise, *Energy and Empire,* 1989, pp. 587–589, 647.

21. Smith and Wise, *Energy and Empire,* 1989, p. 58.

22. Buchanan, *The Engineers,* 1989, pp. 169–171.

23. Report of the Committee appointed to inquire into construction of submarine cables, *Parliamentary Papers,* 1860, *62,* p. v; and for the chemistry of gutta-percha insulation, p. vii and appendix 4.

24. Smith and Wise, *Energy and Empire,* 1989, p. 25.

25. Robert Louis Stevenson, "Memoir," chapter 3, in *Papers Literary, Scientific &c. by the late Fleeming Jenkin,* ed. Sidney Colvin and J. A. Ewing, 2 vols., London: Longmans, Green 1887, volume I; also Smith and Wise, *Energy and Empire,* 1989, p. 698.

26. D. S. Landes, *The Unbound Prometheus,* 1969, p. 235.

27. J. D. Scott, *Siemens Brothers, 1858–1958: an essay in the history of industry,* London: Weidenfeld & Nicolson, 1958, pp. 46, 51. Important improvements to the dynamo were made almost immediately by Gramme (a Belgian) and Altenek (an engineer employed by Siemens and Halske).

28. Peter Temin, *Iron and Steel in Nineteenth-Century America,* Cambridge, MA, MIT Press, 1964, pp. 128–136. I am indebted to Alison Armstrong for advice on dolomite furnace linings.

29. The German, Rudolf Clausius, came to similar conclusions to those of Thomson a year earlier. See D. S. L. Cardwell, *From Watt to Clausius,* 1971.

30. Smith and Wise, *Energy and Empire,* 1989, p. 525.

31. Charles Babbage, 'Introduction', in Peter Barlow, *A Treatise on the Manufactures and Machinery of Great Britain,* London, 1836, p. 81.

32. Smith and Wise, *Energy and Empire,* 1989, p. 658.

33. W. Stanley Jevons, *The Coal Question: an Inquiry concerning the Progress of the Nation and the probable exhaustion of our Coal Mines,* London and Cambridge, 1865, p. 215.

34. Jevons, *The Coal Question,* 1865 ed., pp. 344–349 (Jevons's italics); also see *The Coal Question,* 3rd ed., edited by A. W. Flux, London: Macmillan, 1906, frontispiece.

35. Louis C. Hunter, *A History of Industrial Power in the United States, 1780–1930, Volume 1, Water Power,* Charlottesville: University of Virginia Press, 1979.

36. See the useful discussion of electric, steam, and gasoline-engined automobiles in George Basalla, *The Evolution of Technology,* 1988, pp. 198–204.

37. Basalla, *The Evolution of Technology,* 1988, p. 70.

38. C. H. Gibbs-Smith, *Aviation: an Historical Survey,* 1985 ed., p. 64. A new biography of the Wrights gives further perspectives; see Tom Crouch, *The Bishop's Boys: a life of Wilbur and Orville Wright,* London: Nelson, 1990.

39. David A. Hounshell, *From the American System to Mass Production,* 1984, pp. 193–202, 214.

40. Elting E. Morison, *From Know-how to Nowhere,* 1974, pp. 126, 130–131.

41. W. H. McNeill, *The Pursuit of Power,* 1982, pp. 280–282. For American research on continuous-aim firing of naval guns, see Elting E. Morison, *Men, Machines and Modern Times,* Cambridge, MA, MIT Press, 1966, pp. 17–44.

42. McNeill, *The Pursuit of Power,* pp. 292–293.

43. Avner Offer, *The First World War: an agrarian interpretation,* Oxford: Clarendon Press, 1989, pp. 226–227. On Tirpitz's plans for the German navy, and the "naval arms race" with Britain, see pp. 323–325.

44. Susan J. Douglas, "The Navy adopts the radio, 1899–1919." In Merritt Roe Smith (ed.), *Military Enterprise and Technological Change,* Cambridge, MA: MIT Press, 1985, pp. 117–174; see also Offer, *The First World War,* 1989, pp. 329–330, 349.

45. The significance of this is brought out well by Basalla, *The Evolution of Technology,* 1988, p. 158.

46. John B. Rae, *Climb to Greatness: the American aircraft industry, 1920–60,* Cambridge, MA: MIT Press, 1968.

10. Idealistic Trends in Twentieth-Century Technology

On idealism (and fantasy) in modern technology, there are useful discussions by Basalla, Florman, Noble, also Winner.

1. David A. Hounshell, *From the American System to Mass Production,* p. 260.

2. Le Corbusier, *Vers une architecture,* Paris: Editions Cres, 1923; trans. Frederick Etchells and John Rodker, London: Architectural Press, 1946; on the general context, see Nikolaus Pevsner, *Pioneers of Modern Design,* Pelican ed., Harmondsworth: Penguin, 1960.

3. Hounshell, *From the American System to Mass Production,* pp. 311–314.

4. Maxwell Fry, *Fine Building,* London: Faber, 1944, pp. 76–77.

5. Hounshell, *From the American System to Mass Production,* p. 308.

6. J. D. Bernal, *The Social Function of Science,* London: Routledge, 1939, pp. 190–191, 383.

7. Bernal, *The Social Function of Science,* pp. 148–149.

8. Maurice Goldsmith, *Sage: a Life of J. D. Bernal,* London: Hutchinson, 1980, p. 98.

9. Solly Zuckerman, *From Apes to Warlords,* London: Hamish Hamilton, 1978, p. 404.

10. Goldsmith, *Sage: a Life of J. D. Bernal,* p. 141.

11. Gary Werskey, *The Visible College,* London: Allen Lane, 1978.

12. Robert Olby, *The Path to the Double Helix,* London and Basingstoke: Macmillan, 1974, p. 268.

13. The standard history is: R. G. Hewlett and O. E. Anderson, *The New World, 1939–1946: A History of the U.S. Atomic Energy Commission,* vol. 1, University Park, PA: Pennsylvania State University Press, 1962.

14. Hewlett and Anderson, *The New World,* pp. 232–239.

15. Robert Jungk, *Brighter than a Thousand Suns: a Personal History of the Atomic Scientists,* trans. James Cleugh, London: Gollancz, 1958; new ed., Harmondsworth: Penguin, 1960, pp. 124–125.

16. F. I. Ordway and M.R. Sharpe, *The Rocket Team,* London: Heinemann, 1979, pp. 42, 47.

17. Elting E. Morison, *From Know-how to Nowhere,* 1974, pp. 120–146.

18. Werskey, *The Visible College,* pp. 220–221.

19. Herbert F. York, *The Advisors: Oppenheimer, Teller and the Superbomb,* San Francisco: W. H. Freeman, 1976.

20. Freeman J. Dyson, *Disturbing the Universe,* New York: Harper & Row, 1979.

21. Don K. Price, *Government and Science,* New York: New York University Press, 1953, new ed., New York: Galaxy Books, 1962, p. 1.

22. Hewlett and Anderson, *The New World,* pp. 1–2.

23. Price, *Government and Science,* pp. 48–49, quoting Vannevar Bush, *Science—the Endless Frontier: a report to the President,* Washington, D.C.: Government Printing Office, 1945, p. 181.

24. Price, *Government and Science,* p. 52.

25. Hewlett and Anderson, *The New World,* p. 619.

26. Peter Pringle and James Spigelman, *The Nuclear Barons,* New York: Holt, Rinehart and Winston, 1981, pp. 158–160.

27. Barry Russell, *Building Systems, Industrialization and Architecture,* London and New York: John Wiley, 1981, pp. 441–449.

28. Goldsmith, *Sage: a life of J. D. Bernal,* pp. 141–142; Hilary Rose and Steven Rose, *Science and Society,* Harmondsworth, Pelican, 1970, pp. 91–93.

29. Russell, *Building Systems,* pp. 454–455, 472.

30. Alison Ravetz, *The Government of Space: town planning in modern society,* London: Faber, 1986, p. 84. See also Hugh Griffiths, Alfred Pugsley, and Owen Saunders, *Report of the Inquiry into the Collapse of Flats at Ronan Point,* London: Her Majesty's Stationary Office, 1968.

31. Fry, *Fine Building,* p. 63.

32. C. H. Waddington, *The Scientific Attitude,* Harmondsworth: Penguin, 1941, pp. 124–125.

33. Cecil O. Smith, "The longest run: public engineers and planning in France," *American Historical Review, 95,* 1990, see pp. 689–692.

34. Hilary Rose and Steven Rose, *Science and Society,* Harmondsworth: Penguin, 1970, p. 158.

35. Solly Zuckerman has described this process in a range of publications: "Science advisers and scientific advisers," *Proceedings of the American Philosophical Society, 124,* 1980, pp. 241–255, reprinted as "The deterrent illusion," *The Times* (London), 21 January 1980; also *Nuclear Illusion and Reality,* London: Collins, 1982; see Herbert York's comments also, in York, *The Advisors,* pp. ix, 81.

36. Bruce Mazlish (ed.), *The Railroad and the Space Program,* Cambridge, MA: MIT Press, 1965, pp. 8, 40–41.

37. Jacques Ellul, *The Technological Society,* trans. John Wilkinson, New York: Vintage Books, 1964, pp. 9–10.

38. David Elliott and Ruth Elliott, *The Control of Technology,* London and Winchester: Wykeham, 1976, especially chapter 5, "Social Control of Technology."

39. Walter C. Patterson, *Nuclear Power,* Harmondsworth: Penguin 1976, p. 232; see also Pringle and Spigelman, *The Nuclear Barons,* on an "ice age" in nuclear technology. On the SST, see Basalla, *The Evolution of Technology,* pp. 189–190.

40. Witold Rybczynski, *Paper Heroes: a Review of Appropriate Technology,* New York: Anchor Books, and Dorchester (England): Prism Press, 1980, pp. 87–92.

41. See Schumacher's essay, "Buddhist economics," written in 1966, in E. F. Schumacher, *Small is Beautiful,* London: Blond and Briggs, 1973, pp. 48–56.

42. For a fuller discussion of insights derived from feminism, see Pacey, *The Culture of Technology,* chapter 6. On domestic equipment and western women, see Ruth Schwartz Cowan, "The industrial revolution in the home: household technology and social change in the 20th century," *Technology and Culture, 17,* 1976, pp. 1–22. On approaches to housing comparable with intermediate technology, I am advised that one of the best books is Christopher Alexander, *The Production of Houses,* New York: Oxford University Press, 1985.

43. Godfrey Boyle (ed.), *A.T. in the Eighties,* Milton Keynes: Open University Appropriate Technology Group, 1986; see Peter Harper on origins of "alternative technology," pp. 56–57.

44. Wil Lepkowski, "Wanted: new science policy," *Nature* (London), *258,* 1975, pp. 374–375; also Daniel Deudney and Christopher Flavin, *Renewable Energy: the Power to Choose,* New York: W. W. Norton (for Worldwatch Institute), 1983, pp. 142, 204–207.

45. *The Global 2000 Report to the President: Entering the Twenty-First Century,* Washington, D.C.: Government Printing Office, 1980.

46. Deudney and Flavin, *Renewable Energy,* pp. 204, 212, 264, 291.

47. David F. Noble, "Automation madness," in Steven L. Goldman (ed.), *Science, Technology and Social Progress,* Bethlehem: Lehigh University Press, 1989, pp. 74–76. See also, William J. Broad, *Star Warriors,* New York: Simon and Schuster, 1985.

48. Brooke Hindle, *Emulation and Invention,* 1981; also C. S. Smith, "Art, technology and science," *Technology and Culture, 11,* 1970, pp. 493–549.

49. Dennis Gabor, *Innovations: Scientific, Technological and Social,* New York: Oxford University Press, 1970, p. 9.

50. Basalla, *The Evolution of Technology,* pp. 71, 78.

51. Elting E. Morison, *From Know-how to Nowhere,* especially chapter 9.

52. T. S. Eliot, *Choruses from "The Rock," III, Collected Poems 1909–1962,* London: Faber, 1963, p. 169.

53. P. B. Medawar, *The Hope of Progress,* London: Methuen, 1972, pp. 125, 127.

54. J. R. Ravetz, *Scientific Knowledge and its Social Problems,* London: Oxford University Press, 1971, pp. 434–436.

55. Francis Bacon, *Novum Organum,* 1620, book I, aphorism 124; also very similar wording in *The Advancement of Learning,* 1605, book I, 5, 11; see Arthur Johnston (ed.), *The Advancement of Learning and New Atlantis,* by Francis Bacon, Oxford: Clarendon Press, 1974.

Select Bibliography

This bibliography lists books of general interest only, many of them relevant as background for several chapters in this book. Specialist books and articles in academic journals are covered in the notes for each chapter.

Adams, F. D., *The Birth and Development of the Geological Sciences,* London: Baillière, 1938; new ed., New York: Dover, 1954.

Agricola, Georgius, *De re metallica,* translated from the Latin ed. of 1556 by Herbert Clark Hoover and Lou Henry Hoover, *Mining Magazine,* 1912; new ed., New York: Dover, c. 1950.

Aitchison, Leslie, *A History of Metals,* 2 vols., London: Macdonald & Evans, 1960.

Artz, F. B., *The Development of Technical Education in France, 1500–1800,* Cambridge, MA: MIT Press, 1966.

Ashton, T. S., *Iron and Steel in the Industrial Revolution,* Manchester: Manchester University Press, 1963.

Bagrow, Leo, *A History of Cartography* (ed. R. A. Skelton), London: Watts, 1964.

Basalla, George, *The Evolution of Technology,* Cambridge and New York: Cambridge University Press, 1988.

Binnie, G. W., *Early Victorian Water Engineers,* London: Thomas Telford, 1981.

Branner, Robert, *Burgundian Gothic Architecture,* London: Zwemmer, 1960.

Brown, R. A., Colvin, H. M., and Taylor, A. J., *The History of the King's Works,* 2 vols., London: Her Majesty's Stationary Office, 1963.

Buchanan, R. A., *The Engineers: a History of the Engineering Profession in Britain, 1750–1914,* London: Jessica Kingsley, 1989.

Bud, Robert, and Roberts, Gerrylynn K., *Science versus Practice: Chemistry in Victorian Britain,* Manchester: Manchester University Press, 1984.

Burstall, Aubrey F., *A History of Mechanical Engineering,* London: Faber, 1963.

Cardwell, D. S. L., *From Watt to Clausius: the Rise of Thermodynamics in the Early Industrial Age,* London: Heinemann, 1971.

Chambers, J. D., and Mingay, G. E., *The Agricultural Revolution,* London: Batsford, 1966.

Cipolla, Carlo M., *European Culture and Overseas Expansion,* Harmondsworth: Pelican, 1970 (incorporating Cipolla's earlier works, *Clocks and Culture,* London: Collins, 1967, and *Guns and Sails,* London: Collins, 1966).

Cobban, Alfred, *A History of Modern France,* vol. 1, *The Old Regime and the Revolution, 1715–1799,* London: Cape, 1962.

Cole, C. W., *Colbert and a Century of French Mercantilism,* 2 vols., London: Cass, 1964.

Condit, C. W., *American Building Art: the Nineteenth Century,* New York: Oxford University Press, 1960.

Crombie, A. C., *Augustine to Galileo: Science in the Middle Ages,* London: Heinemann, 1957; 2nd ed., Heinemann, 1961.

Deane, Phyllis, and Cole, W. A., *British Economic Growth 1688–1959,* Cambridge: Cambridge University Press, 1962.

Dickinson, H. W., *A Short History of the Steam Engine,* Cambridge: Cambridge University Press, 1938; new ed. with introduction by A. E. Musson, London: Cass, 1963.

Donald, M. B., *Elizabethan Copper: the History of the Company of Mines Royal,* London: Pergamon, 1955.

Drake, Stillman, *Discoveries and Opinions of Galileo,* New York: Doubleday, 1957.

Dunsheath, Percy, *A History of Electrical Engineering,* London: Faber, 1962.

Finer, S. E., *The Life and Times of Sir Edwin Chadwick,* London: Methuen, 1952, reprinted 1970.

Fitchen, John, *The Construction of the Gothic Cathedrals,* London: Oxford University Press, 1961.

Fitton, R. S., *The Arkwrights: Spinners of Fortune,* Manchester: Manchester University Press, 1989.

Fitton, R. S., and Wadsworth, A. P., *The Strutts and the Arkwrights,* Manchester: Manchester University Press, 1958.

Florman, Samuel C., *The Existential Pleasures of Engineering,* New York: St. Martin's Press, 1976.

Frankl, Paul, *The Gothic—Literary Sources and Interpretations,* Princeton, N.J.: Princeton University Press, 1960.

Gibbs-Smith, Charles H., *Aviation: an Historical Survey to the End of World War II,* London: Her Majesty's Stationary Office, 1970, 2nd ed., 1985.

Gille, Bertrand, *The Renaissance Engineers,* London: Lund Humphries, 1966.

Gimpel, Jean, *The Cathedral Builders,* trans. C. F. Barnes, New York: Evergreen Profile Books, 1961.

Gimpel, Jean, *The Medieval Machine: the Industrial Revolution of the Middle Ages,* London: Victor Gollancz, 1977.

Haber, L. F., *The Chemical Industry during the Nineteenth Century,* Oxford: Clarendon Press, 1958.

Hall, A. Rupert, *The Scientific Revolution, 1500–1800,* London: Longman, 1962.

Harvey, John, *The Medieval Architect,* London: Wayland, 1972.

Hill, Donald, *A History of Engineering in Classical and Medieval Times,* London: Croom Helm, 1984.

Hills, R. L., *Power in the Industrial Revolution,* Manchester: Manchester University Press, 1970.

Hills, R. L., *Richard Arkwright and Cotton Spinning,* London: Priory Press, 1973.

Hills, R. L., *The Stationary Steam Engine,* Cambridge: Cambridge University Press, 1989.

Hindle, Brooke, *Emulation and Invention,* New York: New York University Press, 1981.

Hounshell, David A., *From the American System to Mass Production 1800–1932,* Baltimore: Johns Hopkins University Press, 1984.

Hunter, Louis C., *Steamboats on the Western Rivers: an Economic and Technological History,* Cambridge, MA: Harvard University Press, 1949.

Hyde, C. K., *Technological Change in the British Iron Industry,* Princeton, N.J.: Princeton University Press, 1977.

Jensen, Martin, *Civil Engineering around 1700,* Copenhagen: Danish Technical Press, 1969.

Kearney, Hugh, *Science and Change, 1500–1700,* London: Weidenfeld & Nicolson, 1971.

Keller, A. G., *A Theatre of Machines,* London: Chapman and Hall, 1964.

Klemm, Friedrich, *A History of Western Technology,* trans. Dorothea Waley Singer, London: Allen & Unwin, 1959.

Klingender, Francis D., *Art and the Industrial Revolution,* ed. Arthur Elton, London: Paladin, 1972.

Landes, D. S., *The Unbound Prometheus: Technological Change and Industrial Development in Western Europe from 1750,* Cambridge: Cambridge University Press, 1969.

Mack, Mary, *Jeremy Bentham: an Odyssey of Ideas 1748–1792,* London: Heinemann, 1962.

McNeill, W. H., *The Pursuit of Power: Technology, Armed Force and Society since AD 1000,* Chicago: Chicago University Press, 1982; Oxford: Basil Blackwell, 1983.

Mazlish, Bruce (ed.), *The Railroad and the Space Program,* Cambridge, MA: MIT Press, 1965.

Merchant, Carolyn, *The Death of Nature: Women, Ecology and the Scientific Revolution,* New York: Harper and Row, 1980.

Mitchell, B. R., with Deane, Phyllis, *Abstract of British Historical Statistics,* Cambridge: Cambridge University Press, 1962.

Morgan, B. G., *Canonic Design in English Medieval Architecture,* Liverpool: Liverpool University Press, 1961.

Morison, Elting E., *From Know-how to Nowhere: the Development of American Technology,* New York: Basic Books, and Oxford: Blackwell, 1974.

Morrison, Philip, and Morrison, Emily (eds.), *Charles Babbage and his Calculating Engines: Selected Writings,* New York: Dover, 1961.

Mumford, Lewis, *Technics and Civilization,* London: Routledge, 1946.

Musson, A. E., and Robinson, Eric, *Science and Technology in the Industrial Revolution,* Manchester: Manchester University Press, 1969.

Needham, Joseph, Wang Ling, and Price, Derek J. de Solla, *Heavenly Clockwork: the Great Astronomical Clocks of Medieval China,* Cambridge: Cambridge University Press, 1960.

Nef, John, *The Conquest of the Material World,* Chicago: Chicago University Press, 1964.

Noble, David F., "Automation madness," in Steven L. Goldman (ed.), *Science, Technology and Social Progress,* Bethlehem, PA: Lehigh University Press, 1989.

Pacey, A., *The Culture of Technology,* Cambridge, MA: MIT Press, 1983.

Pacey, A., *Technology in World Civilization: a Thousand-year History,* Cambridge, MA: MIT Press, 1990.

Panofsky, Erwin, *Abbot Suger on the Abbey Church of St. Denis,* Princeton, N.J.: Princeton University Press, 1946.

Parsons, W. B., *Engineers and Engineering in the Renaissance,* Baltimore, 1939; new ed., R. S. Woodbury (ed.), Cambridge, MA: MIT Press, 1967.

Pevsner, Nikolaus, *An Outline of European Architecture,* Harmondsworth: Pelican, 1943; 5th ed., 1957.

Price, D. J. de Solla, "On the origin of clockwork," *Contributions from the Museum of History and Technology,* Washington, D.C.: Smithsonian Institution, 1959, pp. 81–112.

Rich, E. E., and Wilson, C. H., (eds.), *Cambridge Economic History of Europe,* vol. 4, Cambridge: Cambridge University Press, 1967.

Robinson, Eric, and Musson, A. E., *James Watt and the Steam Revolution,* London: Adams and Dart, 1969.

Rolt, L. T. C., *George and Robert Stephenson: the Railway Revolution,* London: Longman, 1960.

Rolt, L. T. C., *Tools for the Job: a Short History of Machine Tools,* London: Batsford, 1965.

Rolt, L. T. C., *Victorian Engineering,* London: Allen Lane, 1970.

Rosenberg, Nathan, *Technology and American Economic Growth,* New York: Harper & Row, 1972.

Rossiter, Margaret W., *The Emergence of Agricultural Science: Justus Liebig and the Americans 1840–1880,* New Haven, CT: Yale University Press, 1975.

Saul, S. B., (ed.), *Technological Change: the United States and Britain in the Nineteenth Century,* London: Methuen, 1970.

Simson, Otto von, *The Gothic Cathedral,* London: Routledge, 1956; 2nd ed., New York: Bollingen, 1962.

Singer, C., Holmyard, E. J., Hall, A. R., and Williams, T. I., (eds.), *A History of Technology,* 7 vol., Oxford: Clarendon Press, 1954–59 and 1978. (Vols. 6 and 7, 1978, ed. T. I. Williams.)

Smith, C. T., *An Historical Geography of Western Europe before 1800,* London: Longman, 1967.

Smith, Crosbie, and Wise, M. Norton, *Energy and Empire: a biographical study of Lord Kelvin,* Cambridge: Cambridge University Press, 1989.

Smith, Cyril Stanley, *A History of Metallography,* Chicago: Chicago University Press, 1960.

Smith, Merritt Roe (ed.), *Military Enterprise and Technological Change,* Cambridge, MA: MIT Press, 1985.

Smith, Norman, *A History of Dams,* London: Peter Davies, 1971; Secaucus, N.J.: Citadel Press, 1972.

Steinman, David B., and Watson, Sara Ruth, *Bridges and their Builders,* New York: Putnam, 1941.

Tawney, R. H., *Religion and the Rise of Capitalism,* London: Murray, 1926; Harmondsworth: Penguin, 1964.

Temin, Peter, *Iron and Steel in Nineteenth-Century America: an Economic Inquiry,* Cambridge, MA: MIT Press, 1964.

Trinder, Barrie, *The Industrial Revolution in Shropshire,* Chichester, England: Phillimore, 1973.

Webster, Charles, *Samuel Hartlib and the Advancement of Learning,* Cambridge: Cambridge University Press, 1970.

Webster, Charles, *The Great Instauration: Science, Medicine and Reform 1626–1660,* London: Duckworth, 1975.

Webster, Charles, *From Paracelsus to Newton: Magic and the Making of Modern Science,* Cambridge: Cambridge University Press, 1982.

White, Lynn, *Medieval Technology and Social Change,* London: Oxford University Press, 1962.

Winner, Langdon, "Technological frontiers and human integrity," in Steven L. Goldman (ed.), *Science, Technology and Social Progress,* Bethlehem, PA: Lehigh University Press, 1989.

Wittkower, Rudolf, *Architectural Principles in the Age of Humanism,* 3rd ed., London: Tiranti, 1962.

Wolf, A., *A History of Science, Technology and Philosophy in the 16th and 17th Centuries,* 2 vols., London: Allen & Unwin, 2nd ed., 1950–52.

Index

Note: Page numbers in *italics* refer to illustrations and their captions. (The index was compiled with the help of Honor Farrell.)